CONCEPTS IN SOLIDS

CONCEPTS IN SOLIDS

Lectures on the Theory of Solids

P. W. Anderson

Princeton University

World Scientific

Singapore • New Jersey • London • Hong Kong

Published by

World Scientific Publishing Co. Pte. Ltd.
P O Box 128, Farrer Road, Singapore 912805
USA office: Suite 1B, 1060 Main Street, River Edge, NJ 07661
UK office: 57 Shelton Street, Covent Garden, London WC2H 9HE

British Library Cataloguing-in-Publication Data
A catalogue record for this book is available from the British Library.

CONCEPTS IN SOLIDS
Lectures on the Theory of Solids

ISBN 981-02-3195-4
ISBN 981-02-3231-4 (pbk)

Printed in Singapore by Uto-Print

Vita

P. W. Anderson

Philip W. Anderson has been the Joseph Henry Professor of Physics at Princeton University since 1978. Professor Anderson received his Ph.D. in physics from Harvard University in 1949. He received the Nobel Prize in Physics in 1977 for work in condensed matter physics. He has taught at Cambridge University (1967-1975) and has been a Fellow and Honorary Fellow at Jesus College (1969-1975) and a Visiting Fellow at Churchill College, Cambridge University (1961-1962). From 1949 to 1984 Dr. Anderson served at Bell Laboratories as Chairman of the Theoretical Physics Department (1959-1961), Assistant Director of the Physical Research Laboratory (1974-1976) and as Consulting Director of the Physical Research Laboratory (1976-1984). In addition to the Nobel Prize, Professor Anderson's honors include the Oliver E. Buckley Prize of the APS (1964), the Dannie Heineman Prize of the Academy of Science at Gottinger (1975), the Guthrie Medal and Prize (1978), the National Medal of Science (1983), the Foreign Association Academia Lincei (1985) and an Honorary Fellowship from the Institute of Physics (1985). He is a member of the American Philosophical Society and has been a Foreign Fellow of the Indian Academy of Sciences, a Foreign Fellow of the Japan Academy of Sciences, a Foreign Member to the Royal Society and has been a member of the American Academy of Arts and Sciences and the National Academy of Sciences Council. He has lectured at the University of Wisconsin, Duke University, the University of California at San Diego, and Harvard University.

Special Preface

I reread *Concepts in Solids* with both pride and embarrassment. Pride, both because it was this set of lectures which inspired Brian Josephson to invent his effect--not every book can point to the precise Nobel prize it inspired--and because I did, in a very brief space, manage to touch some of the key topics which are still not adequately covered in your average solid state theory book. For instance, it is shocking that the main texts used in this country still do not touch on the Mott transition or the "Magnetic State." I was aiming at conceptual, not mechanical physics, and I hope I got there.

Embarrassment, because after all, there has been 30 years of physics since then. For instance, I note that I guessed absolutely wrong in dismissing tight-binding theory out of hand: it has not yet totally come into its own but it is, in my present opinion, the right way to think about most bonding in solids. I am not ashamed of skipping localization--only Mott was interested in it, and neither of us yet knew where to go next. I was prescient about broken symmetry—as Josephson realized—but left out phase transitions, as I myself noted.

Nonetheless, I believe that the average student will still be harmed less by this book than by any number of other books I should not name, and I welcome the reissuance.

P. W. Anderson
March 1991

Preface

These notes were for a course given at the Cavendish Laboratory, Cambridge, in the fall and winter terms of 1961-1962. Nominally, it was for second- and third-year graduate students who had had a survey course in solid-state physics, and were interested (at least) in theory; but I assumed very little formal theoretical background. I think the notes can be read by anyone who has had a thorough course in quantum mechanics, but the reader who knows something about solids will find them much easier, and will also not be misguided by my rather arbitrary and specialized choice of material.

The idea of the course was to teach a number of central concepts of solid-state physics, trying to choose those — band theory, nearly free electrons, effective Hamiltonian theory, elementary excitations, broken symmetry — which lay as near as possible to what I consider to be the main stream of development of the subject. Such a choice is necessarily arbitrary — whose fields, such as dislocation theory, transport theory and fluctuation-dissipation theorems, magnetic resonance theory in all its forms, and critical fluctuations, which could easily be argued to be quite as important, were ignored, simply because the course was of finite length. My choice of examples was even more arbitrary. For instance, the choices of the electric field case of effective Hamiltonian theory and of excitons to illustrate collective excitations were made because I thought the students were likely to encounter the more usual examples elsewhere. From time to time, to liven the course up a bit, I introduced original material; the discussions of the limitations of nearly free electron theory, of the philosophies of elementary excitation theory, and of broken symmetry are new, and that of the magnetic state is not widely available.

The language and presentation are very informal; very few changes were made from my original lecture notes as written. I might add that little effort has been made to bring them up to date. Both limitations are of course implicit in the idea of a lecture note volume.

I would like to express my gratitude for the hospitality of Professor Mott and the

Cavendish Laboratory, the secretarial staff of which prepared the original version of the notes. Mr. Liu Sham was kind enough to edit the notes and write in most of the equations.

P.W. ANDERSON

Murray Hill, New Jersey
January 1963

Contents

Concepts in Solids

1

INTRODUCTION

A. PREPARATION AND TEXTS

The subject of these lecture notes is "Concepts in the Theory of Solids"— in point of fact I should have said "in the *Quantum* Theory of Solids," because there is very little of our understanding of the properties of matter which does not depend to some extent on the quantum theory. Some acquaintance with the quantum theory will be necessary: a certain understanding of the matrix formalism and transformation theory as well as of elementary wave mechanics. For instance, time-dependent perturbation theory and operator equations of motion will be used without much further explanation but not techniques of modern field theory; these will be, if necessary, derived from scratch.

What preparation in the solid-state area is necessary is determined by the intent of this course, which is not to survey the phenomenology of the properties of solids but to go somewhat more deeply into what is behind these properties. In many cases, this means that we shall try to understand *why* solids behave as they do, but in many others of course—perhaps more—we shall simply be coming to the questions at which our real understanding fails. It will then, clearly, be a great help to have a reasonably wide knowledge of what the properties of solids are. Kittel's "Introduction to Solid State Physics" (1) is an excellent text which surveys the field on a level preparatory to what will be said. In other words, some familiarity with such concepts as Debye T^3 specific heat, Brillouin zones, free-electron specific heat or spin paramagnetism, electron or nuclear paramagnetic resonance, and others of the more or less standard theoretical ideas and experimental techniques will be assumed. No texts have even attempted to cover solid-state theory as a whole at any basic level since Seitz in 1940 completed "Modern Theory of Solids" (2); and that is in fact by far the best text still. This may indicate that we have not made much progress since 1940 in basic understanding, only in investigating much wider classes of phenomena, which is to some extent true. In any case, the only answer so far found to the problem of modernizing the "Modern Theory" has been to issue a series of books containing review articles, the so-called

"Seitz-schrift" or Seitz-Turnbull series (Solid State Physics—Advances in Research and Applications), which is probably the best single source at this point. Kittel's book is very good, but except possibly in the one field of crystal symmetry in the second edition, it is not very complete on any one subject. Special areas are covered reasonably well in certain books—e.g., Ziman's "Electrons and Phonons" (3) and other books on special subjects such as dislocations, magnetic resonance, etc. Wannier's "Elements of Solid State Theory" (4) and Peierls' book (5) should be mentioned, each of which is probably the result of some such selection as will be made here, but a quite different one. A magnificent, but quite advanced, and quite condensed text is Landau and Lifshitz' "Statistical Physics" (6). A forthcoming text which will cover a very wide area of solid-state physics from the point of view of group theory and symmetry principles will be M. Lax, "Symmetry Principles in Solid-State Physics" (6a). In any case, books and articles used for sources in any given area will be mentioned.

B. PLAN OF THE COURSE

1. First some general ideas about solid-state physics, including the classification of solids into types, and some broadly general things about the quantum chemical facts we might hope to explain from a deeper point of view: occurrence, properties, etc., of the different types.

2. The next and possibly most important and fundamental area is the purely one-electron band theory, since probably most of the basic questions such as binding, symmetry, band structures, etc., are primarily determined by the bands.

What we hope to lead up to, via some study of the older methods and results, are the more recent ideas of Phillips, Cohen, Heine, and others about the success of the almost-free electron model, and some of the speculations one can then make about binding-energy trends and the quantum chemistry of solids.

Another subject will be the modern developments in one-electron band theory in the presence of perturbations, external fields, impurities, etc.: i.e., effective mass theory.

3. Next we comment on the reason why such manifest oversimplifications as the one-electron theory discussed above work—namely, the idea of elementary excitations, probably the single most fruitful theoretical concept in all of solid-state physics. After discussing the theoretically simple case of insulators and the less simple case of metals, we go on to discuss various possible kinds of collective elementary excitation—excitons, spin waves, and phonons. There are also some more general remarks about collective excitations which apply to all of these.

4. In discussing spin waves we shall treat magnetism. Here the fundamental questions are why and when materials are magnetic—the question of the magnetic state, or when free spins occur—and what causes the interactions among the spins which lead to ferro- or antiferromagnetism. We shall also use this as an example of some general facts about condensation.

C. GENERALITIES AND CLASSIFICATION OF SOLIDS

I suppose any course on the theory of solids should start with a definition of a solid, although we all know that the physicist rejects the definition of a solid as (roughly) what hurts your toe when you kick it, and defines it as a regular array of atoms in the sense of having, to a good approximation, translational symmetry under some one of the space groups.

At this point we have already slipped past three of the most fundamental questions of solid-state physics: (1) Why is a solid? (2) How does one describe a solid from a really fundamental point of view in which the atomic nuclei themselves as well as the electrons are treated truly quantum-mechanically? (3) How and why does the solid hold itself together?

The first of these can be the subject of bitter semantic and philosophical argument, but as far as I know no one can give a real proof, or even a good qualitative reason, as to why the ground or lowest energy states of almost all assemblages of atoms are regular rather than irregular in nature. For instance, in the simplest possible case of a box full of rigid spheres under pressure, everyone assumes for obvious reasons that the regular closest packing (cubic or hexagonal) has the smallest volume and therefore, under pressure, the lowest energy; but I know of no proof that that is so.

Landau has expressed this point of view (7): that the ground state of such a system must have some invariance group. After all, the initial Hamiltonian H_{el} + H_{nuc} is invariant under the full space translation and rotation group. In one case we know about—He_4 at low pressure—and possibly another—He_3—the ground state of the system as a whole retains the full translation and rotation group, although in the latter it may be only full translation symmetry that is maintained. In all other cases, the system condenses, by which we mean it chooses a still lower symmetry, namely a periodic space group. It is quite unreasonable in fact to expect a system starting from a Hamiltonian with the full translation-rotation group symmetry to have a lowest state with no symmetry at all. Perhaps it is a more interesting question why there are not more quantum liquids, not why there are no cases of glass-like lowest energy states. I feel this is an interesting point of view but hardly a convincing argument.

The second question may be one which present interest in solid He_3 and He_4 will see solved in at least an approximate way fairly soon, although I have never yet seen a satisfactory fully quantum-mechanical description of a solid, with all the zero-point motion adequately included, starting from a realistic description of the atomic interactions.

As far as the third question is concerned—how and why a solid holds itself together—I think we will give an adequate if approximate account when we come to talk about phonons and collective excitations.

Knowing then, as an experimental fact, that solids do exist, we can ask in general what sorts of solids there are and how one might classify them. There are a number of phenomenological ways of classifying solids—for instance, one which you may find most familiar is by symmetry, which is a rigorous mathematical way of going about it, and a most useful one, but not directly related to the binding forces and other physical properties of solids.

In Seitz' and Kittel's opening remarks one finds a classification which is much more to the point of this course, according to the type of chemical bond which, from a phenomenological point of view, is responsible for the binding energy of the substances. Seitz' classification contained five categories: (1) metals, (2) ionic crystals, (3) valence or covalent crystals, (4) molecular crystals, and (5) semiconductors.

By now we realize that in every real sense the distinction between semiconductors and metals o, valence crystals as to type of binding, and between semiconductors and any other type of insulator as to conductivity, is entirely artificial; semiconductors do not represent in any real sense a distinct class of crystal. To the remaining four categories Kittel added a fifth, of great interest from the point of view of dielectrics and ferroelectrics but otherwise not distinct in any very important way from molecular or ionic crystals: hydrogen-bonded crystals. Thus we have a reasonable classification into 5 types (see Table 1).

I note in connection with each type of crystal the most significant properties, which can in almost all cases be shown to follow from ideas about the forces which bind the crystals. I hope these connections may become clear to you in the course of these notes. Finally, it is significant to put down as a last column the areas of the periodic table in which each category occurs.

There is one noteworthy thing about this table of occurrences which is not discussed very often. If we look simply at the elements on the left- and right-hand sides of the periodic table: the elements, say Li and Na, with one extra electron in the s and p bands, or even at Al with one p electron, as opposed to those elements with one, two or even three holes in the p bands, we find that the former are metals where the latter tend to form molecular or at best valence crystals. One is so often led to believe in a fundamental symmetry between holes and electrons that it is worthwhile to point out that chemically they are quite dissimilar. Later on, in discussing magnetism, we shall find quite the same dichotomy between the behavior of a few holes in the d shell as opposed to a few electrons. The two types of phenomena may be related and will be discussed later.

It is also important to discuss the extent to which this classification fails. Naturally one can think immediately of a number of obvious cases in which the classification fails, but the principle still holds — e.g., the ammonium halides, $(NH_4)^+Cl^-$, etc.; here the NH_4^+ is molecularly bound, the crystal as a whole ionic with a slight overtone of hydrogen bonding. The occurrence of molecule-like groups in ionic crystals is quite common.

More important and interesting by far is the existence of a range of substances, which, it is by now understood, completely fill in the territory between the three strongly bound categories of metals, ionic and valence crystals. For instance, let us start from NaCl, which is certainly almost purely ionic in nature, and increase the valency of the two constituents in the same row of the periodic table. MgS still remains in the typically ionic NaCl structure, and obeys most of the relationships one expects for an ionic crystal. AlP, on the other hand, is already a member of the group of semiconducting valence-like crystals of ZnS structure (zincblende) of III-V elements; more completely investigated are the closely analogous InP and AlSb. We all know that Si itself is a valence semiconductor par excellence. With increasing atomic number the true ionic crystals become rarer and rarer; ZnS

TABLE 1

Type	Type or "canonical" case	Cause of binding	Properties	Occurrence
Molecular crystals	Ne or N_2	Van der Waals attraction: Mutual polarization, an n-body effect	Resistivity R high. Binding weak. n^2(refr. index), ϵ (diel. const.) low: electrons seldom mobile. Close-packed structures	(a) rare gases (b) lower, right-side elements: C, N, O, H, F, S, P, C1, etc. *No* electropositive, *few* heavy elements
Hydrogen-bonded crystals	H_2O	Hydrogen bonds — lowering of K. E. of proton by 0–H—O	Moderately weak binding (~10-100 x molecular); n^2 low, peculiar dielectric properties because of mobile p. Loose structures	Same as above except always contain H and an electronegative element · O, F etc.
Ionic crystals	NaC1	Ionic bond: Madelung potential + shell effects which make ionization energies right	Very strong binding, close-packed structures (rather), R high, n^2 low, ϵ high (ionic pol.), ionic conduction, low-mobility electronic conduction	Metal from left (electropositive) side—Na, K, Ca—with a right-side, preferably low at.-wt. element—O, F, C1
Metals	Na, Al	Metallic bond—low K.E. of electrons because of free motion in metal	Moderate to strong binding, close packing, R low, $dR/dT>0$ (metallic conduction). Metallic appearance (=low-frequency absorption band)	Widest occurrence: to left of H, B, Si, As, Te; if no atoms to right it is a metal; it cften is anyway. Combinations of elements at random usually metals
Valence crystals	Diamond	"Valence bond"— like bond of organic chemistry. Distinct in principle from metal?	Hard, strongly bound, loose-packed structures. High n^2 and ϵ . R variable, but high mobility conductivity	Rare but important. Low Z, central elements: C, N, P, Si, B are typical constituents

itself, CdS, CdTe are all very valence- or even metal-like in character. Ca, Sr, and Ba, on the other hand, tend to retain their property of forming ionic-like oxides and sulfides—although one could explain this on the basis that the next orbitals available to them, those which are filled in their neighbors Sc and V, are d and not p orbitals and so are unsuitable for tetrahedral bonding—e.g., in terms of a valence crystal argument. It has been shown quantitatively that most of the alkali halides are pretty good ionic crystals in the sense that the charges on the two ions are pretty much ± 1, but otherwise the quantitative degree to which this is so, even in oxides, is very much in doubt. The silicates, for instance, typically crystallize in valence-like, not ionic, structures.

Even more disturbing is the fact that the distinctions between valence crystals and metals and between ionic crystals and metals are gradually losing their clear-cut character with our increasing knowledge of the broad and little-investigated field of intermetallic compounds. I cannot hope to give you a reasonable insight into this area because there are few clear ideas as to the quantum chemistry behind the bewildering variety of phenomena. One example will show how bad things can be, namely, the stoichiometric intermetallic compound NaTl(8). This is a metal with good conduction, not a semimetal, of a structure such that the Tl's form a diamond lattice, the Na's occupying the large interstices in the Tl lattice which themselves also form a diamond lattice. Since the Tl is such a relatively large ion, this is by no means a close-packed structure, in fact it is one to be expected of a typically valence-type crystal. The only way one can make sense out of that is to suggest that the Na's have ionized to Na^+, donating the odd electron to Tl which is now Tl^-, having an sp^3 configuration suitable for the diamond lattice. The fact that the Na is indeed ionized is confirmed by nuclear resonance evidence, the details of which I shall not discuss here. Thus we have in a single substance metallic conduction of a reasonable order, ionic charge-transfer, and an open, valence-type structure.

This then is a brief survey of the general features of the quantum chemistry of solids. I would repeat that I have made no attempt at completeness, even relative to Seitz' first chapter, in which he discussed, for instance, order-disorder, atomic-size relationships, and the Hume-Rothery alloys, subjects I pass over only because of lack of time, and because they are well treated in the literature, not because of their lack of interest and importance. I would emphasize that a whole area, that of intermetallic compounds, remains almost unexplored.

Another way in which the lines, at least between valence crystals and metals, have begun to blur is the general realization that the basic electronic structure of a valence crystal like Si, which can be a fairly good insulator, is not noticeably different, when looked at in a broad sense, from the electronic structure of a good metal. That is, when one studies the electrons with a tool such as plasma resonance or positron annihilation, which resolves electronic density distributions and the like only on the scale of volts—i.e., tenths of rydbergs—rather than tenths of volts, they resemble quite closely the free electron Fermi sphere, as is also true of many good metals. It is to be emphasized that the nature of the binding forces in such crystals has as yet not been amenable to calculation, so that we do not have a clear quantitative idea as to the sources of the binding energy and whether they are more valence- than metallic-like.

A few other ways of categorizing solids might be discussed broadly here. There is, for example, the question of their magnetic properties. The great majority of all solids are either diamagnetic or, in the case of a fairly large number of metals, slightly paramagnetic with little temperature dependence. There are, however, groups of substances in which there is evidence that the atoms in the solid state retain free, orientable spin and in some cases orbital magnetic moments. This may be evidenced by the presence of strongly temperature-dependent paramagnetism, increasing at low temperatures, or by the phenomenon of ferro- or antiferromagnetism—i.e., ordered arrangements of magnetic moments, occurring most often at reasonably low temperatures.

The widest occurrence of such moments is perhaps in the more or less ionic salts of atoms with unfilled shells of 3d, 4f, or 5f (or to a lesser extent 4d and 5d) electrons - the iron, rare earth, and actinide groups. The canonical example is MnF_2. It is a fair generalization that these are electrons which are in inner shells, in the sense that the bulk of their density is to be found inside the last maximum in the density of the outer, valence electrons. The occurrence or nonoccurrence of such moments in ionic crystals follows fairly well-known and well-understood rules of the quantum chemistry of ionic complexes.

Another category, the magnetic molecular crystals, solid crystals of organic free radicals, is, while of considerable interest to chemists, probably only worth a footnote here.

On the other hand, in crystals which are not so ionic—oxides and sulfides of earlier members of the 3d group such as Ti and V, or of 4d elements—the situation is by no means as clear-cut, as Morin has shown in a series of investigations (9), and the boundary between the magnetic and the nonmagnetic state is still being drawn and is not completely understood.

The situation in metals is even more confused. Here, aside from the f groups, in all but two known cases only substances containing one of the later members of the iron group —Mn through Ni—are ever magnetic. There are a large number of compounds (e.g., $CoSi_2$, almost the first superconducting compound of nonsuperconducting elements to be found) of magnetic elements which are not magnetic, but so far only two metals, $ZrZn_2$ and $ScIn_3$, not containing magnetic ingredients, are found to be magnetic. I need hardly say that the quantum chemistry of this situation is almost a complete mystery at this writing.

Another phenomenon, superconductivity, seems to give us a broad classification of metallic elements, alloys, and intermetallic compounds,[1] depending relatively little on structural details and in some broad way on the basic quantum chemistry. Superconductivity seems to be the rule rather than the exception among metals. One can in fact state the requirements for its occurrence thus: Practically the only islands of nonsuperconductivity among nonmagnetic metals occur at

[1]As an alloy I will hereafter define a substance with some perceptible range of composition variation and thus presumably at least a partially random atomic arrangement; as a compound, a substance with a well-defined stoichiometry and atomic arrangement. It is convenient not to use the word alloy for the latter.

very small valence-electron to atom ratios — normally two or less and around a few elements in the sixth column — W (no longer Mo) and alloys of similar valence to atom ratio. Magnetic metals and semimetals are also not superconducting. Again the reasons for these rules are at best only qualitatively understood (10).

2

ONE-ELECTRON THEORY

A. HARTREE-FOCK THEORY

1. General Philosophy of Hartree-Fock

Naturally the starting point and the basis for the one-electron theory are the Hartree-Fock equations, which play an absolutely fundamental role in solid-state physics. On Hartree-Fock theory good sources are Seitz' chapter and appendix, and Reitz' article in the Seitzschrift (11). One may get the impression that modern many-body theory, of which one hears so much, goes far beyond Hartree-Fock, and that therefore we should not bother with such old-fashioned stuff. This is not at all true—in fact, modern many-body theory has mostly just served to show us how, where, and when to use Hartree-Fock theory and how flexible and useful a technique it can be. For instance, Cohen and Ehrenreich (12), among others, have related plasma and correlation energy to Hartree-Fock; Valatin, Thouless (13), and others have studied nuclear collective motion by a "time-dependent" Hartree-Fock theory; and even the theory of superconductivity can be brought into a form almost identical with Hartree-Fock (14). In magnetism, the biggest and most important recent developments are the achievements of the "unrestricted" Hartree-Fock method (15).

The basic idea of the Hartree-Fock equations is this: One wishes to find the lowest eigenstate $\Psi(r_1 \ldots r_n)$ of a system of interacting electrons having a hamiltonian

$$H = \sum_j \left[\frac{P_j^2}{2m} + V(r_j)\right] + \frac{1}{2} \sum_{jk} \frac{e^2}{|r_j - r_k|}$$

where $V(r_j)$ is an external potential of some sort — the potential of the nucleus in an atomic problem, of the atom cores, usually, in a solid-state problem, etc.

It is difficult, however, to solve problems involving more than one electron in an external potential—even He and H_2, the two simplest many-electron problems, are susceptible only to a rather lengthy numerical analysis which shows no capability at all of generalization to more complicated systems. The first thing one hopes to do, then, is to find some way of treating one electron at a time. The obvious way to do this is to treat the electrons as statistically independent: that is, to use a product wave function:

$$\Psi(r_1, r_2, \ldots, r_N) = \psi_1(r_1)\psi_2(r_2)\ldots\psi_N(r_N)$$

since in this way the probability density $|\psi|^2$, as well as all other single-electron quantities of each electron, are independent of those of the others. One would then try to calculate each of the functions by solving a wave equation in which the interaction e^2/r_{ij} is replaced by its average value over the probability distribution of all the other electrons:

$$\left(\frac{P_i^2}{2m} + V(r_i)\right)\psi_i + \sum_{j\neq i}\int dr_j\, |\psi_j(r_j)|^2 \frac{e^2}{|r_i - r_j|}\psi_i(r_i) = E_i\psi_i(r_i)$$

This is the Hartree approximation; it was later proved that these equations assured that the resulting total energy should be an extremum relative to all neighboring product-type functions. As you can see, the only way to solve these equations is self-consistently. One must assume a set of ψ_j, recalculate the ψ's using the mean interaction term derived from the assumed ψ_j's, and substitute these again in the Hartree equations, finally hoping to make the result self-consistent.

The product function, however, does not satisfy the Pauli exclusion principle, nor the more fundamental restriction of the Fermi statistics that the wave function should be antisymmetrical; in fact, the antisymmetric wave function which comes closest to obeying the hypothesis of statistical independence of all possible one-electron quantities such as probability density, current, etc., is the Slater-Fock determinant,

$$\Psi(r_1 \ldots r_N) = (N!)^{-1/2}\begin{vmatrix} \psi_1(r_1) & \ldots & \psi_1(r_N) \\ \psi_2(r_1) & \ldots & \psi_2(r_N) \\ \vdots & & \vdots \\ \psi_N(r_1) & \ldots & \end{vmatrix}$$

(*Note:* The coordinate r_j is here assumed to denote both space and spin variables.)

In this determinantal wave function Ψ, on the other hand, the motions of pairs of electrons are not entirely uncorrelated, because, for instance, the exclusion principle requires that if $r_1 = r_2$ and the spin coordinates of the two electrons are also identical, then the first two columns are identical and the determinant vanishes. I'll leave it as an exercise for the reader to show that in the case of just two electrons the probability density of finding one at r_1, the other at r_2 is

$$1/2 \{ |\psi_1(r_1)|^2 \, |\psi_2(r_2)|^2 + |\psi_1(r_2)\psi_2(r_1)|^2$$
$$- [\psi_1^*(r_1)\,\psi_2(r_1)\,\psi_1(r_2)\,\psi_2(r_2) + \text{c.c.}]\}$$

(if the spins are parallel; if antiparallel, the second term is absent).

Clearly this changes the mean interaction e^2/r_{12} which should be used to determine the wave function ψ_1, and Fock showed from the variational principle that the new correct wave equation is the Hartree-Fock equation:

$$-\frac{\hbar^2}{2m}\nabla_1^2\,\psi_{i\sigma}(r_1) + V(r_1)\psi_{i\sigma}(r_1) + \sum_{j,\sigma'}\left[e^2\int\frac{|\psi_{j\sigma'}(r_2)|^2}{|r_1 - r_2|}\,dr_2\right]\psi_{i\sigma}(r_1)$$

$$-\left(\sum_j e^2\int\frac{\psi_{j\sigma}^*(r_2)\psi_{i\sigma}(r_2)}{|r_1 - r_2|}\,dr_2\right)\psi_{j\sigma}(r_1) = E_i\psi_{i\sigma}(r_1)$$

Note that our sums over j can include $j = i$ because the $j = i$ interaction terms are identical and cancel. Thus the linear operator on the left-hand side is identical for all wave functions, and the single-particle wave functions are automatically orthogonal for different eigenvalues E_i (and of course orthogonalizable if the eigenvalues are the same).

A brief comment on the spin problem. In the first place, a fact that has tended to be forgotten, although some realization of it existed in the literature of the thirties, is that there is, of course, no restriction requiring that the wave-function sets for opposite spins be identical, orthogonal, or have any particular relationship. A solution often exists with identical orthogonal sets of functions, for up and down spins, and can be reasonably accepted in many problems not involving magnetism; but there is no guarantee that it is the best solution even then. In magnetic problems the Hartree-Fock equations for up and for down spins are necessarily different; the biggest difference, and one often neglected, can be seen if we ask what would happen if we calculated the Hartree-Fock energy of the function $\psi_{i,-\sigma}$ if that is empty, but the identical function with spin up, $\psi_{i,\sigma}$, is full. The rather large i = j term, which cancels if $\sigma = \sigma'$, is in this case still present, and can have a large effect. The "novel" idea that the two sets of functions for $\sigma \neq \sigma'$ need not be identical nor orthogonal is called the "unrestricted Hartree-Fock" approximation and it has important consequences in the theory of magnetism.

Two cautions about the Hartree-Fock method in general must be given. First, not only are there no theorems whatever about the uniqueness of solutions, but it is obvious in a number of cases that several completely different types of solutions exist, only one of which has any relationship to the actual ground state of the system under consideration. In other words, one can achieve a perfectly self-consistent solution of the Hartree-Fock equations and have it represent, not only not the actual ground state of the system under consideration (as in the case of the free electron gas when superconductivity is possible) but not even by any means the most stable Hartree-Fock solution, nor even an approximate exact state. An example is given by the problem of H_2: two protons and two electrons (16). It is obvious that if one has a one-electron potential V(r) which has a certain symmetry —as in this case (Figure 1) in the interchange of the two protons—one may try

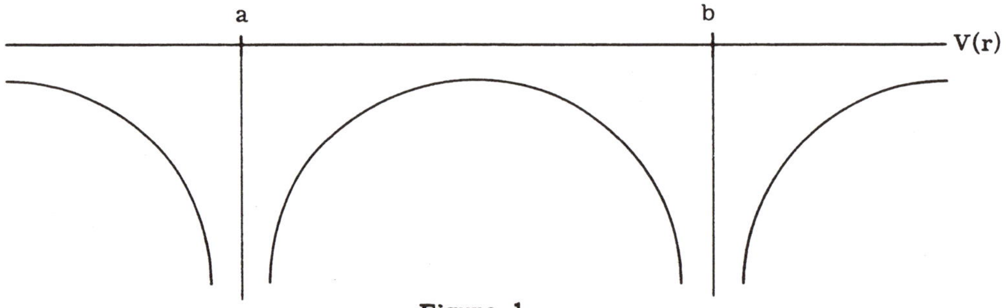

Figure 1

one-electron functions which reflect this symmetry. In this case, if we choose a function $\phi_a(r)$ centered about a, such functions are $\phi_a + \phi_b$ and $\phi_a - \phi_b$, which are even and odd functions, respectively. Now there are various ways in which the self-consistent field may be forced to maintain precisely the initial symmetry of V(r). In this case it is easy: If we put each electron into $(\phi_a + \phi_b)/\sqrt{2} = \phi_{even}$, one with spin up and one with spin down, we clearly retain an even V(r). Another way, under more complicated conditions, would be to fill up all states belonging to a given irreducible representation, i.e., fill up a "shell" or a "band." In both cases we retain the original symmetry—spherical in the atom, periodic in the solid.

The above solution is indeed the correct one, rather close to the ground state of the H_2 molecule, when the protons are close together; we have in fact made up the molecular orbital wave functions for this molecule. But when the protons are far apart, a much superior solution can be obtained by putting one spin-up electron in ϕ_a, the second, spin-down one in ϕ_b. This is the Heitler-London approach, or at least close to it. In the molecular orbital solution, when the molecules are far apart, the two electrons spend half their time on the *same* atom, so that the wave function is, in the separated-atom approximation, half H + H and half $p + H^-$; and the latter energy is far higher than the former. There are true excited states of the system near p + H^-, and true states near H + H, but the M.O. wave function does not have any close resemblance to any real excited state. This is a danger in Hartree-Fock which has been overlooked on occasion—that a perfectly good self-consistent solution may not even be close to any of the actual excited states.

In modern many-body theory tests have been developed (17), which I shall not attempt to discuss, to verify whether a given solution is locally stable relative to other nearby solutions—i.e., whether it is a minimum or a maximum relative to other determinantal functions. This would probably dispose of the difficulty here, but not necessarily so in all cases.

A second point is also illustrated by this example, in the case in which the atoms are far apart. That is, that often a self-consistent solution of the Hartree-Fock equations, which may, as in this case, be the correct one, is nonetheless an unsatisfactory final eigenfunction of the problem, because it does not adequately take the symmetries into account. We have here made up a determinantal wave function with the electron of spin up in wave function ϕ_a, that of spin down in ϕ_b: call it

$$\Psi_{a\uparrow b\downarrow} = \frac{1}{\sqrt{2}} \begin{vmatrix} \phi_a(1)\alpha_1 & \phi_a(2)\alpha_2 \\ \phi_b(1)\beta_1 & \phi_b(2)\beta_2 \end{vmatrix}$$

$$= \phi_a(1)\,\phi_b(2)\,\alpha_1\beta_2 - \phi_a(2)\,\phi_b(1)\beta_1\alpha_2$$

What of $\Psi_{a\downarrow b\uparrow}$? That is, in fact, an entirely equivalent wave function as far as symmetry is concerned; and neither is a satisfactory total wave function because neither transforms according to an irreducible representation of the symmetry group. The correct wave functions are

$$\Psi\ \text{singlet} = \Psi_{a\uparrow b\downarrow} - \Psi_{a\downarrow b\uparrow}$$

$$= \frac{1}{2}\left[\phi_a(1)\phi_b(2) + \phi_a(2)\phi_b(1)\right](\alpha_1\beta_2 - \alpha_2\beta_1)$$

and

$$\Psi\ \text{triplet} = \Psi_{a\uparrow b\downarrow} + \Psi_{a\downarrow b\uparrow}$$

$$= \frac{1}{2}\left[\phi_a(1)\phi_b(2) - \phi_a(2)\phi_b(1)\right](\alpha_1\beta_2 + \alpha_2\beta_1)$$

which is equivalent through a spin rotation to the true Hartree-Fock solution

$$\Psi_{a\uparrow b\uparrow} = \frac{1}{\sqrt{2}} \begin{vmatrix} \phi_a(1)\alpha_1 & \phi_b(1)\alpha_1 \\ \phi_a(2)\alpha_2 & \phi_b(2)\alpha_2 \end{vmatrix}$$

$$= \frac{1}{\sqrt{2}} \alpha_1\alpha_2 \, (\phi_a(1)\, \phi_b(2) - \phi_a(2)\, \phi_b(1)$$

and similarly to $\Psi_{a\downarrow b\downarrow}$.

It is nonetheless possible to obtain from Hartree-Fock not only the triplet's energy but also that of the singlet, if we are willing to go outside the method and rely on our knowledge of spin eigenfunctions. This can be done by comparing the energy of $\Psi_{a\uparrow b\downarrow}$ with that of $\Psi_{a\uparrow b\uparrow}$.

In Hartree-Fock, these two energies differ by a certain term which comes from the additional correlation of the two electrons in the parallel spin case which we derived [Eq. (1)]:

$$\rho(r_1, r_2) = \rho \text{ product } - \Psi_1^*(r_1)\Psi_2^*(r_2)\Psi_2(r_1)\Psi_1(r_2)$$

which leads to an energy difference

$$J = \int dr_1 \;\; dr_2 \;\; \mathcal{H}(r_{12}) \left(\Psi_1^*\Psi_2(r_1) \;\; (\Psi_2^*\Psi_1(r_1) \right)$$

Now the Dirac-Van Vleck vector model tells us that the energies of the states must depend on the spins according to the operator product $J\sigma_a \cdot \sigma_b$, and in the former case the mean value of this is J/4, in the latter -J/4. The true singlet, on the other hand, has energy -3J/4; thus we may estimate its energy by evaluating the exchange integral from the $a\uparrow b\uparrow$ - $a\uparrow b\downarrow$ difference.

This technique has two morals—first, that with a bit of ingenuity one can often go beyond Hartree-Fock quite far, in situations such as spin problems, where we have additional knowledge of the form of the interaction, and where Hartree-Fock alone does not immediately give us the correct symmetry type of eigenfunctions. The same kind of trick is now used in problems such as the rotation of nonspherical nuclei, where Hartree-Fock gives us an ellipsoidal self-consistent field which clearly does not satisfy the necessary requirement that the wave function must be an eigenfunction of angular momentum; but one can quite artificially take the Hartree-Fock solution and rotate it, forcing it into the form of the proper solution (13).

The second moral deals with the same type of problem in an infinite system—e.g., suppose we had discussed a whole lattice of widely separated atoms, as in an antiferromagnet. In such a large system it often occurs that the true solution to the problem is in fact the Hartree-Fock unsymmetric solution rather than a symmetrical substitute—e.g., the true ground state of an antiferromagnet *is* really not a singlet, but rather each atom *has* a definite spin attached to it. But when two Hartree-Fock solutions differ for an *infinite* system, new and very real difficulties come up, because of the peculiar properties of self-consistency. Namely, in order to get from one solution to the other, each of the $N \to \infty$ Hartree-Fock eigenfunctions must undergo a canonical transformation, so that the transformation is in some sense the product of an infinite number of unitary transformations. This means that the two states have essentially zero overlap, the kind of situation which the field theorists make frightening by saying they are actually in different Hilbert spaces. Nonetheless, the fact that in infinite systems two self-consistent solutions can be entirely qualitatively different, as different as two phases—say a solid and a liquid—of the same substance, is of great interest and importance. Similar phenomena occur in superconductivity and even in the everyday question of condensation of a solid—the piece of matter with which we deal can discard a symmetry element of the underlying hamiltonian and set up its own unsymmetric self-consistent field. These problems will be discussed more fully in the final part of these notes.

2. Derivation of Self-Consistent Equations: Second Quantization

All of this has been rather a digression from the application of Hartree-Fock theory to electrons in solids. In order to further emphasize the physics behind Hartree-Fock and to derive a few extra results easily, such as Koopmans' theorem, I shall here give a fairly unusual version of the basic derivation of Hartree-Fock from the variational principle. This will also serve to introduce the concepts of creation and destruction operators which will be of use in later parts of this course. The standard derivation may be found in many texts (see, for instance, Refs. 2 and 11).

As I stated before, practically the only manageable way of representing a many-particle wave function

$$\Psi(r_1 \ldots r_N)$$

is in terms of product functions

$$\phi_1(r_1)\, \phi_2(r_2) \ldots \phi_N(r_N)$$

Of course, this is a very special kind of many-electron function; but it is possible to show that any actual reasonable many-particle wave function may be expanded linearly in terms of a sufficient number of such products—for instance, a many-variable Fourier integral is just such an expansion.

Now a simple product like this is not a suitable wave function for a set of identical quantum mechanical particles, which are in principle not distinguishable. As you all know, the only kinds of particles which appear in nature are either Bose particles, for which symmetric wave functions are required, or Fermi particles, which have only antisymmetric functions. In both cases a suitable generalization of the product exists: for Bose, the symmetrized product

$$(N!)^{-1/2} \sum_P \phi_1(Pr_1)\, \phi_2(Pr_2)\ldots$$

and for Fermi, the antisymmetrized product or determinant

$$(N!)^{-1/2} \sum_P (-1)^P\, \phi_1(Pr_1)\ldots$$

In these wave functions the only relevant information is the fact that a particle is in ϕ_1, another in ϕ_2, etc. If we had chosen the ϕ's from an orthonormal set, in fact, a reasonable way to denote the function would be as

$$\Psi(r_1\ldots r_N) \longleftrightarrow \Psi(n_1 \text{ in } \phi_1, n_2 \text{ in } \phi_2, \ldots) = \Psi(n_1, n_2, n_3\ldots)$$

This is called the "n" or "occupation number representation." For fermions, of course, N may be only zero or 1 by direct computation, while for bosons n may take on any value. Examples:

$$\Psi(n_1 = 1, n_2 = 0, n_3 = 0\ldots) \text{ corresponds to } \phi_1(r_1)$$

$$\Psi(n_1 = 1, n_2 = 1, n_3 = 0\ldots) \longleftrightarrow \frac{1}{\sqrt{2}}\Big(\phi_1(1)\phi_2(2) \pm \phi_1(2)\phi_2(1)\Big)$$

etc.

This can be thought of as simply a linear unitary transformation from one set of variables to another. One way of seeing that this must be so is to take advantage of the freedom of choice of the ϕ to use a particularly simple set, namely, the set of δ functions $\delta(r - R)$, where now the function index n for ϕ_n is the continuous variable R. The function $\Psi(n_R = 1, n_{R'\neq R} = 0)$ is the function with one particle definitely at R; while one may obtain any one-particle function by a unitary transformation

$$\phi(R) \longleftrightarrow \int dR\ \phi(R)\,\Psi(n_R = 1,\ n_{R' \neq R} = 0)$$

which then gives us the unitary transformation

$$\Psi(n_\phi = 1,\ n_{\phi' \neq \phi} = 0) = \int dR\ \phi(R)\ \Psi(n_R = 1,\ n_{R'} = 0)$$

where the wave function $\phi(R)$ plays the role, as in the ordinary one-particle theory, of a unitary transformation $(R \mid \phi)$ between the r-space representation and the ϕ representation. Similar techniques may be employed to express any n-particle function which is properly antisymmetric in terms of the functions labeled by one particle at R_1, a second at R_2, etc. (Hereafter we limit ourselves to fermions.)

The occupation-number representation is thus merely a convenient way of writing an expansion in determinantal wave functions. One begins to obtain formally simpler and more useful results when one introduces further types of quantum-mechanical operators called "creation" and "destruction" operators, operating on these functions in such a way as to add or remove—create or destroy—a particle in a particular wave function.

If we have a set of functions, ϕ_n, which are orthonormalized, we introduce an operator $C^\dagger_{\phi_n}$ or, more simply, $C^\dagger_n$ which, operating on any n-representation function $\Psi(n_1, n_2, \ldots, n_N)$, adds a new particle in the state ϕ_n. That is,

$$C^\dagger_n\ \Psi(n_1, n_2, \ldots, n_n \ldots) = \Psi(n_1, n_2, \ldots, n_n + 1 \ldots)$$

We also introduce its hermitian conjugate C_n, which has precisely the opposite effect, that of reducing n_n by one, i.e., destroying a particle in state n. It is important to note two things: first, that in fact for fermions the determinant with two or more particles in state n vanishes, so that $C^\dagger_n = 0$ if acting on a state Ψ with $n_n = 1$ or greater. This is most easily described by saying $(C^\dagger_n)^2 = (C_n)^2 = 0$. The second is that the phase of the wave function, and thus even its sign, remains undetermined by this definition so far. Thus if we have the wave function

$$\Psi(0 \ldots n_1 = 1, 0 \ldots) \longleftrightarrow \phi_1(r)$$

we don't know yet what phase to give the determinant

$$C^\dagger_{n_2}\ \Psi(0 \ldots n_1 = 1, 0 \ldots) = \Psi(0 \ldots n_1 = 1, n_2 = 1, 0 \ldots)$$

whether to say this is

$$\frac{1}{\sqrt{2}}\left(\phi_1(r_1)\,\phi_2(r_2) - \phi_1(r_2)\,\phi_2(r_1)\right)$$

or vice versa, or to give the corresponding determinant an arbitrary phase factor.

In fact, of course, the phase *is* arbitrary in this case, since physical results cannot depend on it. However, the phase is relevant in comparing two determinants built up by creating particles in various ways—for instance, we may wish to compare the above wave function, which we can write $C_2^\dagger\,(C_1^\dagger\,|\Psi_{vac})$ with one in which we first create the particle in state ϕ_2 and then that in state ϕ_1: $C_1^\dagger\,(C_2^\dagger\,\Psi_{vac})$. Clearly, if the first is written

$$\frac{1}{\sqrt{2}}\,e^{if}\left(\phi_1(r_1)\phi_2(r_2) - \phi_2(r_1)\,\phi_1(r_2)\right)$$

the second must be

$$\frac{1}{\sqrt{2}}\,e^{if}\left(\phi_2(r_1)\phi_1(r_2) - \phi_1\phi_2\right)$$

which is just the negative of the first one. This then may be described by specifying

$$C_1^\dagger\,C_2^\dagger\,(\Psi_{vac}) = -C_2^\dagger\,C_1^\dagger\,|\Psi_{vac})$$

or

$$\left(C_1^\dagger\,C_2^\dagger + C_2^\dagger\,C_1^\dagger\right)|\,\Psi_{vac}) = 0$$

In fact, you can see immediately that in order to make sure that the wave functions remain antisymmetric in interchange of r_1 and r_2, and since the only way we specify which is r_1 and which is r_2 is the order of the factors in the creation product, it is sufficient to make the creation operators for different n *anticommute*: $C_1^\dagger C_2^\dagger + C_2^\dagger C_1^\dagger = 0$, when applied to any wave function.

This anticommutation rule enormously complicates the explicit representation of the $C^\dagger$'s as matrices, so that, although such representations can be found, we shan't give them here.

You can see by taking the Hermitian conjugate that $C_n C_{n'} + C_{n'} C_n = 0$, and that $C_n^2 = C_n^{\dagger 2} = 0$ actually is just the $n = n'$ element of this rule. It is possible to work out for oneself—I leave this as an exercise—that $C_n C^\dagger_{n'} + C^\dagger_{n'} C_n = 0$, $n' \neq n$, also.

This leaves only $C^\dagger_n C_n$ and $C_n C^\dagger_n$. These products are in fact very simple. Consider their effect on $\Psi_0 = \Psi(\ldots n_n = 0, \ldots)$ and on $\Psi_1 = \Psi(\ldots n_n = 1, \ldots)$; the dots represent identical sets of all other indices.

$$C^\dagger_n C_n \Psi_o = 0 \text{ because } C_n \Psi_o = 0$$

$$C_n C^\dagger_n \Psi_1 = 0 \text{ because } C^\dagger_n \Psi_1 = 0$$

$$C^\dagger_n \Psi_o = \pm \Psi_1 \qquad C_n \Psi_1 = \pm \Psi_o$$

The phases are undetermined but they must be the same, so that

$$C_n C^\dagger_n \Psi_o = \Psi_o \qquad C^\dagger_n C_n \Psi_1 = \Psi_1$$

Since all wave functions can be written as linear combinations of these two types, these equations say that for all Ψ,

$$C^\dagger_n C_n + C_n C^\dagger_n = 1; \text{ thus } C^\dagger_n C_{n'} + C_{n'} C^\dagger_n = \delta_{nn'}$$

Also we see that the operator $C^\dagger_n C_n$ is the operator which *measures* the number of electrons in state n.

$$C^\dagger_n C_n = n_n \quad C_n C^\dagger_n = 1 - n_n$$

So far we have shown that we can set up a complete algebra of determinantal wave functions by using the occupation number representation and creation and destruction operators—in fact, we can drop out the former and just start with the vacuum always, writing

$$\Psi(n_1, n_2, \ldots, n_N, \ldots) = (C^\dagger_1)^{n_1} (C^\dagger_2)^{n_2} \ldots (C^\dagger_N)^{n_N} \ldots \Psi_{vac}$$

To make use of this we must also expand physical observables in terms of creation and destruction operators. This is in fact not hard. The way in which it can be done is most simply illustrated by considering the case of a simple one-electron space potential V (r). First let's look at the effect of V(r) on a one-electron function. If we have a set of orthogonal functions $\phi_n(r)$,

$$V(r)\phi_n(r) = \sum_m V_{mn}\phi_n(r)$$

V_{mn} is the usual matrix element $\int \phi_m^* V \phi_n \, dr$. Thus as far as one-electron functions are concerned, V(r) has the same effect as

$$V \longleftrightarrow \sum_{m,n} C_m^\dagger V_{mn} C_n \tag{1}$$

as we see by applying this to $C_n^\dagger \Psi_{vac} \longleftrightarrow \phi_n(r)$.

Similarly, in a two-electron problem, we know that by indistinguishability V(r) must act on both coordinates, so we insert $V(r_1) + V(r_2)$ into the hamiltonian or other observable we study. Calculating the effect of this on a product, $\phi_m(r_1)\phi_{m'}(r_2)$, we get

$$\sum_n \left[\phi_n(r_1)\,\phi_{m'}(r_2)V_{nm} + V_{nm'}\phi_m(r_1)\phi_n(r_2)\right]$$

which is just the effect, again, of destroying either m or m', multiplying by the appropriate V, and replacing by n, i.e., the same operator. (I leave to the reader the exercise of showing that anticommutation of the C's and antisymmetrization lead to the right signs for determinants.)

Now just one more step will make obvious the general rule for forming observables, such as the hamiltonian, in this theory. Let us make the canonical transformation to δ functions δ (r - R), and to the corresponding creation and destruction operators $\Psi^\dagger(R)$, $\Psi(R)$. The notation $\Psi^\dagger$ and Ψ is used here because these operators are often referred to as the *electron field* operators, and play a very similar role to the wave function Ψ of the ordinary Schrödinger equation; they are an operator version of this c-number quantity — hence the name, "second quantization." The transformation is of course

$$C_m^\dagger = \int dR\, \underbrace{\phi_m(R)}_{(m|R)}\, \Psi^\dagger(R) \tag{2}$$

and multiplying through by $(R'|m) = \phi_m^\dagger$ and summing, we get the inverse transformation

$$\sum_m \phi_m^*(R')C_m^\dagger = \int dR \sum_m (R'|m)(m|R)\Psi^\dagger(R)$$

and we recognize the δ function in the sum: $\sum_m (R'|m)(m|R) = \delta(R' - R)$,

$$\Psi^\dagger(R') = \sum_m C_m^\dagger \phi_m^*(R') \tag{3}$$

Inserting Eq. (2) into Eq. (1) we obtain

$$V \longleftrightarrow \sum_{m,n} \int dR \int dR' \ \phi_m(R)\Psi^\dagger(R) \int dR'' \ \phi_m(R'')$$

$$\times \ V(R'')\phi_n(R'')\phi_n^*(R')\ \Psi(R')$$

and again the sums introduce a series of δ functions and we get

$$V \longleftrightarrow \int dR \ \Psi^\dagger(R)V(R)\Psi(R)$$

a formula which looks self-evident. The quantity under the integral sign is often called a "density"— in this case a "potential density."

Again I leave to the reader the proof that the obvious rule can be applied also to the interaction potential

$$\frac{1}{2}\sum_{i,j}\frac{e^2}{R_{ij}} \longleftrightarrow V_{int} = \frac{1}{2}\sum_{m,n,l,p} C_m^\dagger C_n^\dagger C_l C_p V_{mnlp}$$

$$= \frac{1}{2}\sum_{\sigma,\sigma'} \quad dR\, dR' \ \Psi_\sigma^\dagger(R)\Psi_{\sigma'}^\dagger(R') \frac{e^2}{|R - R'|} \Psi_{\sigma'}(R')\Psi_\sigma(R)$$

$$V_{mnlp} = e^2 \int dR \int dR' \ \frac{\phi_m^*(R)\phi_n^*(R')\phi_l(R')\phi_p(R)}{|R - R'|} \quad \delta_{\sigma_m\sigma_p}\delta_{\sigma_n\sigma_l} \tag{4}$$

and, correspondingly, one can write down a kinetic energy density, so that

$$\mathcal{H} = \sum_{\sigma} \int dR\, \Psi_{\sigma}^{\dagger}(R) \left(-\frac{h^2\nabla^2}{2m} + V(R)\right) \Psi_{\sigma}(R) + V_{int} \tag{5}$$

This form in terms of the field operators is useful primarily because it may be expanded in terms of any set of one-electron functions we may wish to use.

Now, at last, we can come to the derivation of the Hartree-Fock equations and Koopmans' theorem in this formalism. Both of these are most easily proved if we take advantage of the freedom of transformation among the various possible sets of φ_n's in terms of which we might express Eq. (5), and express everything in terms of the solutions we hope eventually to achieve. That is, we choose a set φ_n of which the first N are the states occupied in the lowest-energy determinantal wave function. This lowest-energy function is, then, written

$$\Psi_N = \prod_{n=1}^{N} C_n^{\dagger} \Psi_{vac}$$

If Ψ_N is to have the lowest energy, we must require that

$$E_N = \frac{(\Psi_N, \mathcal{H}\Psi_N)}{(\Psi_N, \Psi_N)}$$

be a minimum, and thus that

$$(\delta\Psi_N, \mathcal{H}\Psi_N) = 0$$

where we have ignored the variation of the second Ψ because it gives precisely the complex conjugate, and we shall ensure normalization later.

We can take care of all possible infinitesimal variations of Ψ_N by varying each of the φ_n of which it is composed in turn, that is, by varying each $C_n{}^{\dagger}$. Thus we must ensure that

$$\left(\delta C_n^{\dagger}\left(\prod_{m\neq n}^{N} C_m^{\dagger}\right)\Psi_{vac},\ \mathcal{H} C_n^{\dagger} \prod_{m\neq n}^{N} C_m^{\dagger}\Psi_{vac}\right) = 0$$

If $\delta C_n^\dagger$ is taken as an arbitrary linear combination of $C_m^\dagger$'s: $\delta C_n^\dagger = \Sigma_m \delta_m C_m^\dagger$, we may assume that $\delta_{m=n}$ is zero, to ensure normalization, while those other $m < N$ all give automatically zero because $(C_m^\dagger)^2 = 0$. Thus our testing function $\delta C_n^\dagger = \Sigma_{m>N} \delta_m C_m^\dagger$; we can in effect require: for all $M > N$, $n \leq N$:

$$\left(C_M^\dagger \prod_{m \neq n}^{N} C_m^\dagger \Psi_{vac}, \mathcal{H} \prod_{m}^{N} C_m^\dagger \Psi_{vac} \right) = 0 \tag{6}$$

Straightforward evaluation of this does give the Hartree-Fock equations; but I should like to go by a more physical route and demonstrate Koopmans' theorem at the same time.

Expanding $\mathcal{H}$ in Eq. (5) in terms of our assumed set of solutions, we obtain

$$\mathcal{H} = \sum_{m,n} \mathcal{H}^1_{mn} \; C_m^\dagger C_n + \frac{1}{2} \sum_{lmnp} V_{lmnp} \, C_l^\dagger C_m^\dagger C_n C_p$$

where $\mathcal{H}^1$ is the one-electron part of (5):

$$\mathcal{H}^1_{mn} = \int \varphi_m^*(r) \left(\frac{p^2}{2m} + V \right) \varphi_n(r) \, dr$$

V_{lmnp} has been defined previously as Eq. (4).

Let us now evaluate the *commutator* of $\mathcal{H}$ with the creation operator for one of the occupied states, $C_n^\dagger$. Using our anticommutation relations in a straightforward manner, we can see that

$$\left[C_l^\dagger C_p, C_n^\dagger \right] = C_l^\dagger C_p C_n^\dagger - C_n^\dagger C_l^\dagger C_p = - C_l^\dagger C_n^\dagger C_p - C_n^\dagger C_l^\dagger C_p = 0$$

$$\left[C_n^\dagger C_p, C_n^\dagger \right] = 0$$

$$\left[C_l^\dagger C_n, C_n^\dagger \right] = C_l^\dagger C_n C_n^\dagger + C_l^\dagger C_n^\dagger C_n = C_l^\dagger$$

$$\left[C_n^\dagger C_n, C_n^\dagger \right] = C_n^\dagger$$

The rule is thus that where a product of two operators does not contain C_n it commutes with $C_n^\dagger$; otherwise we annihilate C_n and leave the odd $C_n^\dagger$ operator behind. Thus the commutator with a creation operator has much the same effect as the operator itself. Note that

$$\left[C_a^\dagger C_b, C_c^\dagger\right]^\dagger = -\left[C_b^\dagger C_a, C_c\right]$$

so commutation through a destruction operator leads to the conjugate result with a minus sign.

The commutator of $C_n^\dagger$ through the four-operator product depends in sign on whether C_n appears in the first or second possible place:

$$\left[C_a^\dagger C_b^\dagger C_l C_m, C_n^\dagger\right] = 0 \qquad l, m \neq n$$

$$\left[C_a^\dagger C_b^\dagger C_l C_n, C_n^\dagger\right] = C_a^\dagger C_b^\dagger C_l$$

$$\left[C_a^\dagger C_b^\dagger C_n C_l, C_n^\dagger\right] = -C_a^\dagger C_b^\dagger C_l$$

the last because the C_n must be anticommuted through C_l before annihilating the $C_n^\dagger$.

Using these simple rules we obtain

$$\left[\mathcal{H}, C_n^\dagger\right] = \sum_m \mathcal{H}^1_{mn} C_m^\dagger + \frac{1}{2} \sum_{mlp} (V_{mlpn} - V_{mlnp}) C_m^\dagger C_l^\dagger C_p \tag{7}$$

Now, finally, let us apply this to the evaluation of the energy variation, Eq. (6). We commute any $C_n^\dagger$ through the hamiltonian $\mathcal{H}$; this gives

$$\begin{aligned} 0 = &\left(C_M^\dagger \prod_{m\neq n}^{N} C_m^\dagger \Psi_{vac}, C_n^\dagger \mathcal{H} \prod_{m\neq n} C_m^\dagger \Psi_{vac}\right) \\ &+ \left(C_M^\dagger \prod_{m\neq n}^{N} C_m^\dagger \Psi_{vac}, \left[\mathcal{H}, C_n^\dagger\right] \prod_{m\neq n}^{N} C_m^\dagger \Psi_{vac}\right) \end{aligned} \tag{8}$$

The first term is immediately zero, because state n is definitely occupied in the right-hand state, definitely not in the left. We are left, then, with the scalar product term containing the commutator: the second in Eq. (8) above.

If we did not have the three-fermion terms in $[\mathcal{H}, C_n^\dagger]$, this would be easily solved, simply by demanding that

$$\mathcal{H}^1_{Mn} = 0 \qquad M > N, \quad n \leq N$$

that is, that

$$\left[\mathcal{H}, C_n^\dagger\right] = \sum_{\substack{m \text{ occupied} \\ \text{only}}} \lambda_{mn} C_m^\dagger$$

In fact, the mixing of different occupied states is irrelevant; we could always, by linearly combining the $C_n^\dagger$, diagonalize the matrix λ_{mn} and find a set of states for which

$$\left[\mathcal{H}, C_n^\dagger\right] = E_n C_n^\dagger$$

Precisely the same procedure is followed in principle in determining the effect of the three-fermion terms. A term of the form $C_m^\dagger C_l^\dagger C_p$ will in many cases automatically give zero; p, for instance, must be one of the occupied states, or $C_p \Psi_{vac} = 0$. But if C_p destroys the particle in state j, i.e., $C_p = C_j$, $C_m^\dagger$ or $C_l^\dagger$ must replace it because otherwise the scalar product will be zero; the state on the left has j occupied. Finally, the only way in which M can get occupied is for the other of $C_m^\dagger$ or $C_l^\dagger$ to fill it. We have then the following two types of terms:

$$(V_{Mjjn} - V_{Mjnj})\, C_M^\dagger n_j \qquad \text{and} \qquad -(V_{jMjn} - V_{jMnj})\, C_M^\dagger n_j$$

Looking at the definition of V, we see that it is symmetrical against exchange of inner and outer indices, so that these two are identical and cancel the 1/2 in Eq. (7). Also, we see that $V_{Mnnn} = V_{Mnnn}$, so that we may insert, if we like, the term $C_n^\dagger n_n$, symmetrically. The factor n_j is just unity when applied to the product function, and the final equation is

$$0 = (C^\dagger_M C_n \Psi_{H-F}, \left[\mathcal{H}, C^\dagger_n\right] C_n \Psi_{H-F}) = \mathcal{H}^1_{Mn} + \sum_{m\ occ.} (V_{Mmmn} - V_{Mmnm})$$

or,

$$\sum_{\text{all } m} (\mathcal{H}^1_{mn} C^\dagger_m + \sum_{j=1}^{N} (V_{mjjn} - V_{mjnj}) C^\dagger_m) = E_n C^\dagger_n \tag{9}$$

(Again, we may rediagonalize among the occupied wave functions.)

That this is indeed the Hartree-Fock equation may be verified by transforming to the space (electron field) representation. Equation (9) is then

$$\sum_m \int \phi^\dagger_m(r) \mathcal{H}^1(r) \phi_n(r)\, dr \int dR\, \phi_m(R) \Psi^\dagger(R)$$

$$+ \sum_j \sum_m \int dr \int dr' \frac{e^2}{|r-r'|} \left[\phi^*_m(r)\phi^*_j(r')\phi_j(r')\phi_n(r)\right.$$

$$\left. - \phi^*_m(r)\phi_j(r')\phi_n(r')\phi_j(r)\right] \int dR\ \phi_m(R)\Psi^\dagger(R)$$

$$= E_n \int dR\ \phi_n(R)\Psi^\dagger(R)$$

The sums over m give δ functions in such a way that this becomes

$$\int dR\ \Psi^\dagger(R)\quad \mathcal{H}^1(R)\phi_n(R) + \int dr_2 \frac{e^2}{|r_2-R|} |\phi_j(r_2)|^2 \phi_n(R)$$

$$- \int dr_2 \frac{e^2}{|r_2-R|} \phi^*_j(r_2)\phi_n(r_2)\phi_j(R)$$

$$= \int dr\ \Psi^\dagger(R) E_n \phi_n(R)$$

Since all the $\Psi^\dagger(R)$'s are linearly independent, the coefficient as a function of R must be zero; this coefficient is the Hartree-Fock equation, Q.E.D.

Koopmans' theorem is much easier to prove. It is that the coefficient E_n is precisely the excitation energy required to remove the electron in the state ϕ_n in Hartree-Fock approximation.

The energy of the Hartree-Fock state as a whole, E_{H-F}, may be defined simply as

$$\mathcal{H}\Psi_{H-F} = E_{H-F}\Psi_{H-F} \text{ (+ off-diagonal terms, of course)}$$

The total energy of the state in which electron n is removed is given simply by

$$\mathcal{H} \prod_{m \neq n} C_m^\dagger \Psi_{vac} = \epsilon_n \prod_{m \neq n} C_m^\dagger \Psi_{vac}$$

We may obtain a comparison by using the commutator,

$$C_n^\dagger \mathcal{H} (C_n \Psi_{H-F}) - \mathcal{H} C_n^\dagger (C_n \Psi_{H-F})$$

$$= (\epsilon_n - E_{H-F}) \Psi_{H-F} = - \left[\mathcal{H}, C_n^\dagger \right] C_n \Psi_{H-F}$$

But this last product is precisely what we evaluated in calculating δE; we evaluated the diagonal term, and *defined* it as $E_n C_n^\dagger$; thus

$$\epsilon_n - E_{H-F} = E_n$$

It costs energy E_n to remove one electron in state n. This is Koopmans' theorem.

In fact, in this form it is not quite rigorous because E_{H-F} and ϵ_n are not quite comparable, being the energies of two systems having different numbers of electrons. It can be shown that E_n nevertheless is the meaningful quantity when discussing differences in energy caused by emptying two different states ϕ_n.

This, then, is the physical meaning of the parameters E_n. They are not directly related to the total energy E_{H-F}, because, of course, the E_n contains the interaction energy of the particle n with all the other particles m, and when we remove this particle we lose the whole of this interaction energy. If we then remove a particle m, its removal energy will not contain its interaction with n, although E_m does contain that energy. Thus the sum of all the E_n's must have subtracted from it the total interaction energy in order to avoid counting the interaction twice. Thus

$$E_{H-F} = \sum_n E_n - \frac{1}{2} \sum_{n,m} (V_{nmmn} - V_{nmnm}) \qquad \text{or}$$

$$= \sum_n \mathcal{H}^1_{nn} + \frac{1}{2} \sum_{n,m} (V_{nmmn} - V_{nmnm})$$

The equation $[\mathcal{H}, C_n^\dagger] = E_n C_n^\dagger$ (except for terms which excite triplets of particles) could have been guessed at more or less a priori as a statement of the physical sense of the Hartree-Fock approximation. In other words, what one wishes to do is to find a set of operators which create one-electron excitations, and to fill the lowest N of these one-electron states. The self-consistency requirement on the ground state amounts, then, to asking that the interaction terms, on the average and within the limitations of our approximation, act like an average potential for the states we choose, and not excite other one-electron states—we think of $[\mathcal{H}, C_n^\dagger]$ as $i(\partial C_n^\dagger/\partial t)$ and require that the time development not excite other states. All the complications of manipulation came, not from the calculation of $[\mathcal{H}, C_n^\dagger]$, but in proving that that led to a minimum total energy.

The above is one of the most straightforward examples of the use of commutators with the hamiltonian—i.e., equations of motion—to determine both the ground state and the excitation energies in a many-body problem (18). This is a technique which is useful in most of the branches of many-body theory, although it is probably gradually being replaced by the more general and complete, but similar, method of Green's functions.

B. ENERGY BANDS IN SOLIDS

Now we go on to discuss the much more practical and much less formal questions involved in the use of Hartree-Fock to calculate energy bands in solids.

In this area the articles of Reitz (11), Ham (19), and Wigner and Seitz (20) in Vol. 1 of the Seitzschrift are of great value, as well as Brooks (21) and Jones' book (22). In discussing the calculation of energy bands and cohesive energies, I shall cover the following topics.

1. Perturbation theory for weak periodic potentials—this will give us some concept of Brillouin and Jones zones.

(I will not talk about tight binding, which doesn't seem ever to be a very useful technique in real solids as far as band theory is concerned.)

2. The cellular method.
3. Free electron gas and its exchange energy—brief comments on correlation energy.
4. O.P.W. and cancellation theory.

1. Perturbation Theory for Weak Periodic Potentials. Brillouin and Jones Zones and Symmetrized Plane Waves

No one, of course, knows how to solve exactly— even by machines —the general integrodifferential Hartree-Fock equation in all but the very simplest cases. The approximations that are used are all based simply on estimating some kind of periodic local potential in which the electrons move, and solving the wave equation in this periodic potential $V(r)$. As we shall now discuss, there are various general characteristics common to the solutions of all such wave equations, which are probably most easily described in this, one of the simplest of all approximation schemes, and one which it is now realized applies roughly to many metals.

Of course the simplest of all band structures is that which results in the limit that the periodic potential vanishes entirely. But to understand how to regard that as a band structure, we must first look at the case of a relatively weak periodic potential, $V(r)$.

Every crystal structure has an underlying simple space translation group or "Bravais lattice," composed of translations along three linearly independent vectors $\mathbf{a}_1$, $\mathbf{a}_2$, $\mathbf{a}_3$; and this symmetry ensures that

$$V(\mathbf{r} + l\mathbf{a}_1 + m\mathbf{a}_2 + n\mathbf{a}_3) = V(\mathbf{r})$$

There are always an infinite number of possible ways to choose $\mathbf{a}_1$, $\mathbf{a}_2$, and $\mathbf{a}_3$, but they all generate the same lattice and have the same primitive cell size

$$\Omega = \mathbf{a}_1 \cdot (\mathbf{a}_2 \times \mathbf{a}_3)$$

As a consequence of this periodicity, the Fourier transform of $V(\mathbf{r})$, $V(\mathbf{k})$, reduces to a Fourier sum rather than a Fourier integral:

$$V(\mathbf{k}) = \frac{1}{\text{vol.}} \int d\mathbf{r}\; e^{-i\mathbf{k}\cdot\mathbf{r}} V(\mathbf{r})$$

$$= \frac{n(\text{cells})}{\text{vol.}} \left(\int_{\text{cell}} d\mathbf{r}\; e^{-i\mathbf{k}\cdot\mathbf{r}} V(\mathbf{r}) \right) \sum_{m,l,n} e^{-i\mathbf{k}\cdot(m\mathbf{a}_1 + n\mathbf{a}_2 + l\mathbf{a}_3)}$$

This last sum vanishes when summed over l, m, n unless

$$\mathbf{k}\cdot\mathbf{a}_1 = 2\pi n_1 \quad \mathbf{k}\cdot\mathbf{a}_2 = 2\pi n_2 \quad \mathbf{k}\cdot\mathbf{a}_3 = 2\pi n_3$$

where $n_1 n_2$ and n_3 are all integers. These relationships can only be satisfied when $\mathbf{k}$ is a vector $K = n_1 b_1 + n_2 b_2 + n_3 b_3$ of the *reciprocal lattice* of the Bravais lattice of our crystal. One can define $\mathbf{b}_j$, the basis vectors of the reciprocal lattice, by $\mathbf{a}_i \cdot b_j = 2\delta_{ij}$—i.e., $\mathbf{b}_i$ is perpendicular to two lattice vectors and has a length determined by the third; or by

$$V(r) = \sum_{k} e^{i\mathbf{K}\cdot\mathbf{r}} V(\mathbf{K}) \tag{10}$$

($\mathbf{K}$ is the set of reciprocal lattice vectors.)

This may be inverted to give

$$V(K) = \frac{1}{N\Omega} \int dr\, e^{iK\cdot r} V(r)$$

but, since $e^{iK\cdot(r + la_1 + ma_2 + na_3)} = e^{iK\cdot r}$

$$V(K) = \frac{1}{\Omega} \int_{\text{primitive cell}} dr\, e^{iK\cdot r} V(r)$$

All the consequences of lattice symmetry are summarized in the form of the expression (10).

Now we substitute V(r) into the Schrödinger equation.

$$-\frac{\hbar^2}{2m} \nabla^2 \Psi(r) + V(r)\Psi(r) = E\Psi(r)$$

In the absence of V(r) the solutions will be plane waves,

$$\phi_k(r) = \frac{1}{(N\Omega)^{1/2}} e^{ik\cdot r} \qquad -\frac{\hbar^2}{2m} \nabla^2 \phi_k = -\frac{\hbar^2 k^2}{2m} \phi_k = E_k^o \phi_k$$

The simplest boundary conditions at the surfaces of the crystal which we can impose are the Born-von Karman periodic boundary conditions—that we imagine the lattice folded into itself in each of the three possible directions x, y, and z, so that the lengths in each of the three directions are L_1, L_2, L_3, and $L_1 L_2 L_3 = N$. Thus $\Psi(x + L_1, y, z) = \Psi(x, y, z)$ so that $k_x L_1 = 2\pi n_x$, $k_x = 2\pi n_x / L_1$. Thus there is one possible state for each volume $(2\pi)^3/N\Omega$ in k space, since there is one state for each parallelopiped of volume $(2\pi)^3/L_1 L_2 L_3$ (two states, of course, counting spin).

Now when we introduce V(r) the different plane wave states are no longer independent. We can, however, still hope, if V is weak, that the real wave functions are mostly made up of a single plane wave ϕ_k, and so we label them with a particular k:

$$\Psi_k = \phi_k + \sum_{k' \neq k} a_{k'} \phi_{k'}$$

The Schrödinger equation then becomes the following set of coupled linear equations for the a_k's and E_k,

$$E - E_k^o = \sum_{k' \neq k} (k'|V|k)\, a_{k'} + (k|V|k)$$

$$(E - E_{k'}^o)\, a_{k'} = \sum_{k'' \neq k} (k''|V|k')\, a_{k''} + (k|V|k')$$

We see immediately from the expression for V that in general these matrix elements vanish unless $k' - k$ is some reciprocal lattice vector **K**. The vector $\mathbf{K} = 0$ gives us an average potential $(k|V|k) = V_o$, which is a trivial constant added to all the energies; all the other V's have the effect of coupling **k** with some vector $\mathbf{k} + \mathbf{K}$ connected to it by a translation of the reciprocal lattice. Thus our wave function must really be

$$\Psi_k = \phi_k + \sum_{\substack{K \neq 0 \\ \text{(reciprocal lattice)}}} a_{k+K}\, \phi_{k+K} \tag{11}$$

and our equation system is

$$(E - E_k^o)(1) = \sum_K V_K a_{k+K} \qquad (\text{that is, } a_k = 1)$$

$$(E - E_{k+K}^o)\, a_{k+K} = V_K^*(1) + \sum_{K' \neq -K} V_{K'} a_{k+K+K'}$$

At a general point, perturbation theory can be expected to converge; if so, we see that a_{k+K} is of order V_K, so that the last term on the right is of order V^2 and should be neglected. The result of doing so is

$$a_{k+K} \simeq \frac{V_K^*}{E - E_{k+K}^o}$$

$$E - E_k^o = \sum_K \frac{|V_k|^2}{E - E_{k+K}^o}$$

If, finally, V is sufficiently small, $E \simeq E_k^o$ and we get

$$E_k = E_k^o - \sum_K \frac{|V_k|^2}{E_{k+K}^o - E_k^o} \tag{12}$$

Thus the plane-wave energy spectrum and the plane-wave functions are simply weakly perturbed by an admixture of states of wave vector **k** + **K**, **K** belonging to the reciprocal lattice.

It is worth noting that a wave function of the form of Eq. (11) may be written

$$\Psi_k = e^{i\mathbf{k}\cdot\mathbf{r}} \left(1 + \sum_{K \neq 0} a_{k+K} e^{i\mathbf{K}\cdot\mathbf{r}} \right) = e^{i\mathbf{k}\cdot\mathbf{r}} u_k(r)$$

where, as we saw in expanding V(r), $u_k(r)$ is periodic with the lattice period because it is a Fourier sum over reciprocal lattice wave vectors. This is Bloch's famous result; it obviously does not depend on the convergence of the perturbation theory, but is true in general.

A related fact is that from the point of view of the translational symmetry of the lattice, ϕ_k and ϕ_{k+K} are essentially the same, because if we perform a lattice translation T by $\mathbf{A} = l_1 \mathbf{a}_1 + l_2 \mathbf{a}_2 + l_3 \mathbf{a}_3$,

$$T\phi_k = e^{i\mathbf{k}\cdot\mathbf{A}} \phi_k \qquad T\phi_{k+K} = e^{i\mathbf{k}\cdot\mathbf{A}} \phi_{k+K}$$

Thus group-theoretically they have the same symmetry properties, and can be expected to mix; we shall see the consequences of this when we discuss Brillouin zones.

The expression (12) is not satisfactory when the sum on the right is not small, as will always occur at a large number of places in k space, because there are a number of regions where

$$E_{k+K}^o = \hbar^2 (k + K)^2/2m \simeq \hbar^2 k^2/2m = E_k^o$$

This occurs near points at which

$$2\mathbf{k}\cdot\mathbf{K} + K^2 = 0$$

which occurs whenever the component of **k** || **K** has the value $-1/2\mathbf{K}$, the other two components being arbitrary. Thus, since $-\mathbf{K}$ is a reciprocal lattice vector also, this occurs wherever **k** lies on the plane bisecting a reciprocal lattice vector (Figure 2).

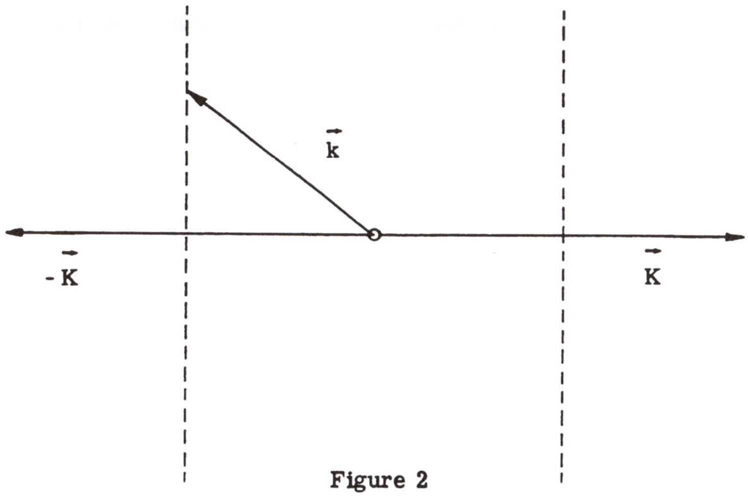

Figure 2

Now if V is really small, and if **k** does not lie at an intersection of two such planes, we can still approximate all the terms in the perturbation sum except the one referring to **K** by using $E_k = E_k^o$. These additional terms then add up to an unimportant constant correction, and we are left merely with a quadratic equation to solve for the two strongly coupled components of the wave function:

$$E - E_k^o = \frac{|V_K|^2}{E - E_{k+K}^o} + \sum_{K' \neq 0, K} \frac{|V_{K'}|^2}{E_k^o - E_{k+K'}^o} .$$

or

$$\left(E - \frac{\hbar^2 k^2}{2m}\right)\left(E - \frac{\hbar^2 (k+K)^2}{2m}\right) \simeq |V_K|^2$$

(where we could, if we had to, correct *both* E_k^o and E_{k+K}^o for the perturbation caused by all the other states. It is not good to correct only one of the two energies, since that causes an unphysical shift in the zone boundary).

Let us set $k = K/2) + k_{\parallel} + k_{\perp}$, $k_{\perp}$ being the irrelevant component perpendicular to K along the bisector plane, and $k_{\parallel}$ that parallel to K, the distance of k from the plane. Then the above equation becomes

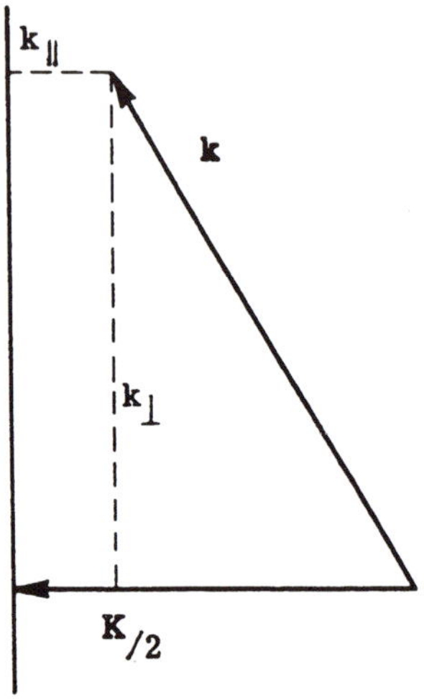

$$\left[E - \frac{\hbar^2}{2m}\left\{(K/2)^2 + k_{\parallel}^2 + k_{\perp}^2\right\} + (\hbar^2/2m)k_{\parallel}K\right]$$

$$\times\left[E - \frac{\hbar^2}{2m}\left\{(K/2)^2 + k_{\perp}^2 + k_{\parallel}^2\right\} - \frac{\hbar^2}{2m}k_{\parallel}K\right]$$

$$= |V_K|^2$$

so that

$$E = \frac{\hbar^2}{2m}\left[(K/2)^2 + k_{\perp}^2 + k_{\parallel}^2\right]$$

$$\pm\sqrt{|V_K|^2 + \left(\frac{\hbar^2}{2m}k_{\parallel}K\right)^2}$$

Figure 3 is a sketch of this equation along a line of constant $K_{\perp}$. This shows the

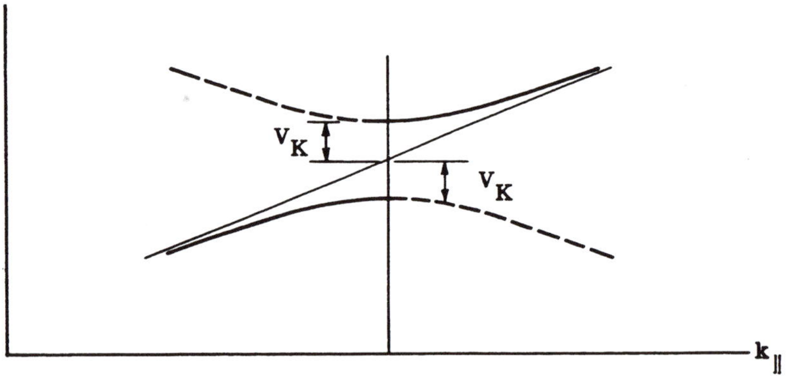

Figure 3

band gap $\Delta_K = 2V_K$ at the Brillouin zone boundary. The periodic potential, therefore, has the effect of breaking the continuity of the E vs. k curve, $E = \hbar^2 k^2/2m$, at each of the planes $(k \cdot K) = |K|^2/2$, so that a sketch of the curve as a whole might look as shown in Figure 4.

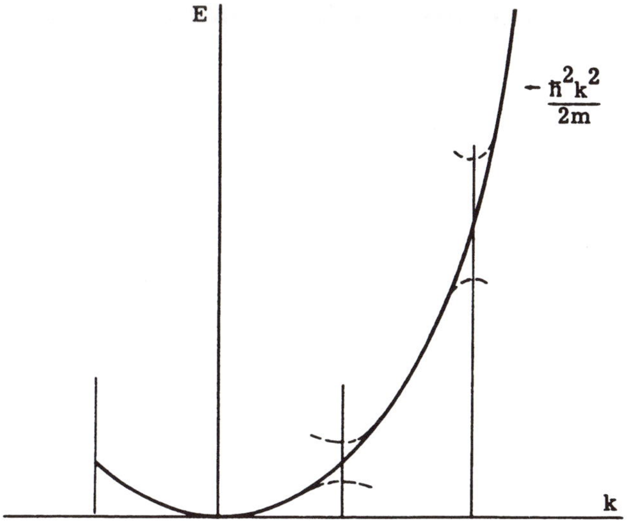

Figure 4

We note that at each of these zone boundary points there is a formal solution of the equation continuing on past the boundary. This formal solution at k is in fact identical with the solution at $k' = k - K$, which is an illustration of a general property that can easily be shown from our starting equation: the energy may be regarded as a continuous but multivalued function of k, with the periodicity of the reciprocal lattice: $E(k + K) = E(k)$.

The planes at which the gaps appear to open are, from that point of view, rather arbitrary; they are picked out by a perturbation calculation, which, in principle, in a sufficiently unsymmetric lattice, does not give any unique prescription—e.g., for where the band gaps are smallest or for where the bands may be flat. Nonetheless the symmetry is usually such that the band structure is even in k space. If it is, the Brillouin zone boundaries can never be really very far from the positions of smallest band separation, and the relative minima and maxima in the energy. A serious violation of this would require strong spin-orbit coupling and the absence of centers of symmetry.

One can show that the innermost set of these plane boundaries encloses exactly one unit cell of reciprocal lattice space—simply because we can draw the cell about each point of R. L. S., and no point can be left out. This is called the *first*

Brillouin zone. We can if we like use the so-called "reduced zone scheme" of translating all energy surfaces by various K's to draw the band structure entirely within this first zone, as a multivalued function of k. The zone clearly has a volume $(2\pi)^3/\Omega$, so it contains exactly N states, one for each primitive cell of the lattice.

Now we go on to the subject of Jones zones or "large zones" (22). So far, I have assumed that the lattice potential V(r) has no symmetry other than the straightforward translation group—I have based everything on the simple Bravais lattice. Many lattices have more than one atom per primitive cell of the Bravais lattice; many have additional rotation or inversion symmetry within the Bravais lattice, e.g., the f.c.c. or b.c.c. Nonetheless, as Jones has shown, this does not necessarily cause any V(K) to vanish. When, however, the lattice symmetry contains screw axes and glide planes in addition to straight translations and rotations, it is possible in many cases for the V(K) to vanish identically for certain sets of planes.

An extremely important example is the diamond lattice (23), which has a glide plane of reflection-translation symmetry. A simple but not rigorous way of finding the extra selection rules on V is to assume the atoms identical and spherically symmetric. As you all know, the Bravais lattice for the diamond lattice is the face-centered cubic lattice with $\mathbf{a}_1 = (a/2,\ a/2,\ 0)$, $\mathbf{a}_2 = (a/2,\ 0,\ a/2)$, etc. The corresponding reciprocal lattice is a b.c.c. one with r.l. vectors $\mathbf{K} = (2\pi/a)(1, 1, 1)$, etc.; (200), etc.; 220; 311; 222; etc.

Now the two atoms in the primitive cell are related by the vector (a/4, a/4, a/4). Thus their contributions to the potential, assuming spherical symmetry, are just proportional to the x-ray structure factor

$$S = 1 + e^{i(a/4,\ a/4,\ a/4)\cdot\mathbf{K}} = 1 + \exp\left\{\frac{i\pi}{2}(l + m + n)\right\}$$

Thus when $l + m + n = 2(2N + 1)$, N an integer, S vanishes: that is, for the 200, 222, 420, 600, etc., planes. According to our perturbation-theory calculation, this would mean that the corresponding energy gap would vanish, and the corresponding Brillouin zone boundaries would disappear.

The fact is that this difficulty can be repaired in a higher order of perturbation theory. In general, even though V_K may vanish there will exist two K's, K_1 and K_2, such that $K_1 + K_2 = K$ and $V_{K_1} \neq 0 \neq V_{K_2}$. For instance, in this case V_{111} and $V_{+1\,-1\,-1}$ do not vanish. Now if k is near the bisector plane of $2\pi/a$ (200), in general E^0_k and $E^0_{k+2\pi/a(111)}$ will not be close, so that we can treat a_{k+K_1} and a_{k+K_2} by perturbation theory. Then

$$(E - E^o_{k+K})\, a_{k+K} = \sum_{K' \neq K} V^*_{K'}\, a_{k+K-K'}$$

$$\simeq V^*_{K_1} a_{k+K_2} + V^*_{K_2} a_{k+K_1}$$

while

$$(E^o_k - E^o_{k+K_1})\, a_{k+K_1} \simeq V_{K_1} a_k$$

so the net effect is to have an *effective* potential for the K zone boundary as in the following equation:

$$(E - E^o_{k+K})\, a_{k+K} \simeq V_K^{\text{eff}}\, a_k$$

$$V_K^{\text{eff}} = \sum_{K_1 + K_2 = K} V_{K_1} \frac{1}{E^o_k - E^o_{k+K_1}} V^*_{K_2}$$

and this effective potential causes an energy gap in precisely the manner in which a real matrix element V_K would.

Thus we have the general result that at every Brillouin zone boundary—except for certain special symmetry points, the discussion of which is quite beyond the scope of these lectures—there is a band gap. Of course this does not mean that as a function of energy there is always a gap, because the gaps at different parts of the zone boundary certainly occur at different energies, so that the bands in general overlap.

In the limit that the potential is weak but not vanishing, we can regain a very simple and often useful approximation to the band structure: namely, neglecting the parts of the bands near the zone boundaries, we look on the free-electron energy $\hbar^2k^2/2m$ as a form of band structure. That is, we take the free-electron paraboloid and fold it back into the reduced zone scheme.

When we regard the free-electron parabola as a kind of band structure, in two or three dimensions rather complicated things happen as we fold each piece of the of the free-electron paraboloid into the first zone; one obtains complicated degeneracies, level places, etc. I reproduce here (Figure 5), for general interest, the resulting energy vs. k curves along the line from the origin to the hexagonal face of the zone boundary for the f.c.c. space lattice, i.e., along $k = (x, x, x)$. As you see, a quite complex band structure in the reduced zone may, when spread out into the "extended zone," be not far from a simple free-electron parabola. Study of those free-electron functions which reflect onto the points of special symmetry in the zone — particularly the origin Γ — allows us to guess from symmetry alone which wave functions at such a point may remain degenerate, and which are only a coincidence of the free-electron approximation. For instance, at the point A all the functions have the form $e^{\pm i(2\pi x/a)} e^{\pm i(2\pi y/a)} e^{\pm i(2\pi z/a)}$ and can be separated out into four groups, or sets of "symmetrized plane waves" (see Figure 5), given in Table 2. These symmetrized plane waves are of importance in band calculations.

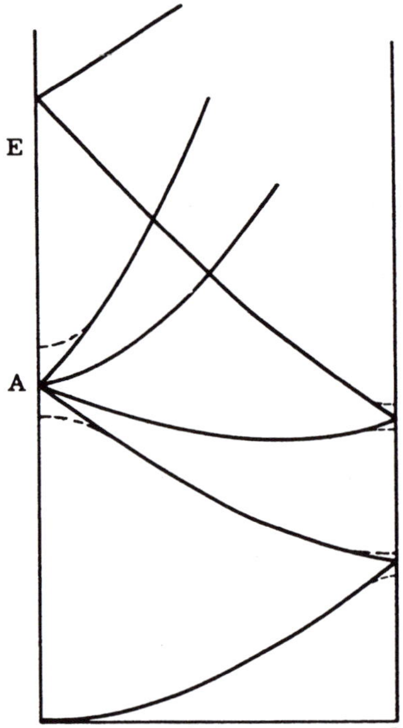

Figure 5

Table 2

Degeneracy	S.P.W.			Label
1:	$\cos \frac{2\pi x}{a}$	$\cos \frac{2\pi y}{a}$	$\cos \frac{2\pi y}{a}$	Γ_1
1:	$\sin \frac{2\pi x}{a}$	$\sin \frac{2\pi y}{a}$	$\sin \frac{2\pi z}{a}$	Γ_2
3:	$\cos \frac{2\pi x}{a}$	$\cos \frac{2\pi y}{a}$	$\sin \frac{2\pi z}{a}$, etc.	Γ_{15}
3:	$\cos \frac{2\pi x}{a}$	$\sin \frac{2\pi y}{a}$	$\sin \frac{2\pi z}{a}$, etc.	Γ'_{25}

Now, to come back to the problem of Jones vs. Brillouin zones, the thing to note is that the gap at a second-order zone boundary, $2V_K^{eff}$, is likely to be considerably smaller than that at a first-order boundary. Thus, while we expect in general a gap at every boundary, there may well be particularly large gaps, in cases with complicated symmetries, only at the true first-order boundaries. These are called the boundaries of the "Jones zones." There is no necessary relationship between the volume of a Jones zone, in terms of number of states per atom, and the volume of a Brillouin zone; the Jones zone need only have some rational number $r > 1$ states of each spin per atom. For instance, again returning to the well-known case of the diamond lattice, the first Brillouin zone is the cell bounded by zone boundaries corresponding mostly to the (111) K vector, but that octahedron has the tips of its six corners cut off by the (200) boundaries. In the Jones zone of diamond, the octahedron's tips are *not* broken off and the cell has 9/8 of an electron per atom.

The search for particularly important zone boundaries may extend beyond the search for boundaries where $V \neq 0$ and on to those for which V is particularly large. In general, we expect such a boundary to be one with a big x-ray structure factor, which can be simply guessed by looking at the x-ray pattern (only of course when all atoms are alike, or nearly so, since the effective potentials for x-rays and low-energy plane electron waves are quite different). For instance, in diamond the structure factor for (111) is

$$F = 1 + \exp\left(\frac{3i}{2}\pi\right) = 1 - i, \quad |F|^2 = 2$$

while that for $l + m + n = 4N$ (220, 440, 400, etc.) is 2, $|F|^2 = 4$. It is then likely that the (220) group of planes bounds a particularly important Jones zone.

If we look at this group of r.l. vectors, we realize that they are the vectors of a face-centered-cubic r.l. of cube-edge size $2\pi/a$ (4), which is therefore the reciprocal lattice of a body-centered lattice of cube size $a/2$, containing eight cubes for every cube in the original f.c.c. lattice. The first *Brillouin* zone of this lattice has, then, since the b.c.c. has only 2 atoms per cube instead of 4, 8/2 or 4 states per atom of the f.c.c. lattice, and 2 per atom of the diamond lattice—4 altogether, counting spin.

Another way to look at this pseudo-body-centered lattice is to draw the diamond lattice, Figure 6. The b.c.c. lattice which gives these strong reflections is

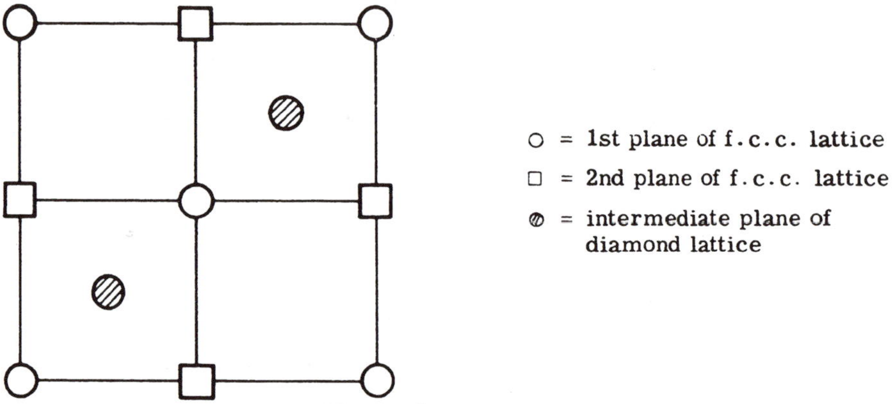

Figure 6

obtained by completing the missing positions of the tetrahedron around each atom to make up a cube, as shown by the O's in Figure 7. This shows that just twice as

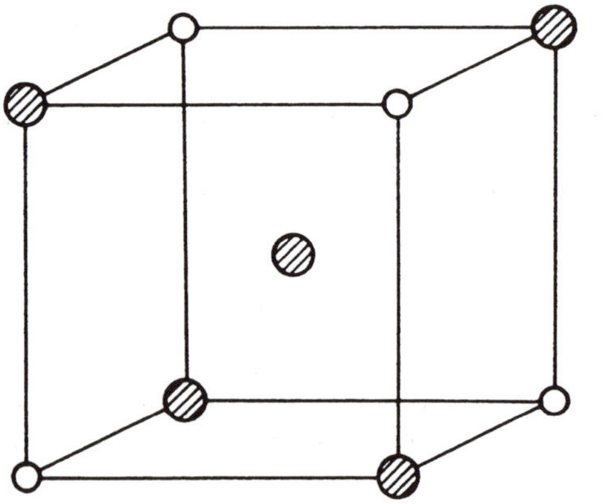

Figure 7

many states fit into its Brillouin zone; there are four electron states per atom in the strong Jones zone of the diamond lattice.

This has all been worked out in detail as a particularly good illustration of the semiempirical validity of the quantum chemical rule which gives the Jones zones their importance: Crystals tend to choose a structure such that a strong, symmetrical Jones zone can be exactly or almost exactly filled. The reason for this is easy enough to see: If the number of electrons in the valence bands of the crystal is such that on the free-electron model the Fermi surface would just come up to the Jones zone boundary, then the development of that boundary lowers the energy of the group of free electrons near the zone boundary, by lowering the energy of occupied electronic states relative to unoccupied ones.

To estimate this effect quantitatively, suppose that the appropriate V_K has magnitude V. On the average, roughly, each electron within energy V of the Fermi surface has had its energy lowered by V/2, so that the lowering is ($V^2/2$ (density of states at Fermi surface). That density is given on the free-electron model as follows:

$$\begin{aligned}\text{density of states per electron} &= \frac{dn}{dE}\,\frac{1}{n}\\ &= \frac{dk}{dE}\left(\frac{1}{n}\,\frac{dn}{dk}\right) = \frac{dk}{dE}\,\frac{3}{k_F}\\ &= \frac{3}{2E_F}\end{aligned}$$

Thus the binding energy per atom would be

$$\sim \frac{4\times 3}{2\times 2E_F}\,V^2 = \frac{3V^2}{E_F}$$

The reader must be cautioned that the numerical factors in this estimate cannot really be evaluated to better than order-of-magnitude accuracy; in fact, the energetics of this situation are in considerable dispute at present (24).

That this can be a rather large effect, and one which is very likely to determine crystal structure, may be seen by estimating its effect in some real substances. For instance, the free-electron Fermi energy in diamond would be 30 ev (about 2 rydbergs), and about half that in the much-less-dense crystals Si and Ge. (This we obtain simply by calculating k_F from $n = k_F^3/3\pi^2$; k_F in diamond turns out to be about $3\times 10^8\ \mathrm{cm}^{-1}$, in Si about $2\times 10^8/\mathrm{cm}$.) Actual values of the components of the potential seen by the free electrons are available from Herman's (23) and Phillips' (25) calculations; they give V_{220} = about 0.4 ryd, or 5 ev in diamond, and about 0.15-0.2 ryd, or 2 ev in Si. Thus the gain in cohesive energy is very large: 3 volts, or half of the total, in diamond, and 1.5 volts in Si. Naturally this is a very rough estimate, but it shows that this can be a real and controlling effect in the usually rather delicate balance of energies which determines the crystal structure a substance will assume. It is an interesting fact that the $\mathbf{K} = 111$ V_K also plays an important role near the band gap in diamond and silicon, being vital in opening up the gap at certain symmetry points at which V_{220} cannot act, especially Γ. Thus a considerable contribution to binding comes also from V_{111}. Probably this is typical of the combinations of several factors which determine observed crystal structures.

There are a number of physical and chemical morals which I should like to draw from this rough calculation before going on to more quantitative things. In the first place, note that we see here a physical reason for the open, valence-type structure quite divorced from the idea of directed bonds as little sticks coming out of the atoms in definite directions. An open structure can represent simply a means of bringing the important Jones zone boundaries out into the area of the Fermi surface in a polyvalent substance. I should, however, caution you that bond directivity does seem to exist as a chemical fact, and that the calculation of which precise structure, having about the right density and Jones zone shape, a given substance will assume, is probably a matter of the complicated interplay of very many Fourier components of V. It usually had better be left to the successful techniques of semiempiricism and chemical instinct for the time being; but we do at least see some of the principles involved, and they do not, as Pauling might suggest, require scrapping the very successful apparatus of the band theory. Second, note that in both cases V was in fact small enough that perturbation theory is not very bad. This is rather a shock for diamond, but true—diamond's gigantic energy gap is just a consequence of the fact that the whole energy scale is wide for carbon, and does not represent a really large departure from the nearly free electron picture.

Finally, I should mention that this application to diamond and silicon is actually one of the most recent such applications; a much older and more thoroughly covered area is the application to certain alloy structures—or rather intermetallic compound structures—and particularly to the famous Hume-Rothery alloys. For a good treatment, see Jones (26). These are compounds of various more-or-less complicated cubic structures, which tend to have a prominent Jones zone of considerable symmetry lying very near the Fermi surface. These are no doubt caused by the same effect, but undoubtedly the energy contribututions are of a considerably smaller magnitude than those in the valence structures. We see this from the fact that these compounds retain their metallic character, indicating that the zone boundary does not open up very much of a gap. It is also worth emphasizing that even, for example, in the diamond case, certain special symmetry directions—the corners of the Jones zone, which are, in terms of symmetry, equivalent to the center of the reduced zone Γ—cannot be split simply by the one component V_{220} in the nearly free picture, so that other components of V, particularly, as we mentioned, V_{111}, play a controlling role in opening up the actual gap which causes insulating behavior.

2. The Cellular Method: Quantitative Calculation of Binding Energy

This is an entirely different method, in many ways, which has shown itself by far the most accurate in calculating cohesive energies of metals. The cellular method, first used by Wigner and Seitz, unfortunately seems to be easily applicable, for known reasons, only to the monovalent alkali metals.

The cellular method contrasts in almost every possible way with the methods based on plane waves. In the almost-free electron model, and, as we shall see, in the O.P.W. method, the wave functions are chosen from the start as proper band functions, satisfying the boundary conditions imposed by the periodicity of the lattice; the effort is then to treat sufficiently accurately, by perturbation or other methods, the effect on these functions of the potentials of the ion cores. The cellular method starts from the attempt to find as exact a potential and as exact a solution as possible in the region of the ion cores, and proceeds to approximate the boundary conditions enforced by periodicity. It appears that in calculating energies this is the more accurate procedure, but that it is only so for the lowest wave functions in the s band of a solid.

The procedure is to divide the real space lattice into unit cells in precisely the same way one forms the Brillouin zone in the reciprocal lattice: by bisecting the vectors to the near neighbors. In the b.c.c. lattice usually assumed by the alkalies, this gives us the regular octahedron of ¼ ¼ ¼ planes chopped off at the corners by ½ 00 planes—a total of 14 faces—with which we are already familiar. The idea of the cellular method is that this polyhedron is not really very different from a sphere, and that in the outer regions in which the departure occurs, the potential is very weak and the wave function smooth. Thus the approximation is made that the unit cell is precisely spherical. This causes a number of simplifications in the problem of numerically integrating the wave equation for the low states:

1. The remainder of the lattice may as well be assumed spherically symmetric also; since the distance between atoms in the alkalies is such that the cores overlap very little, the potential from the other cells is only electrostatic, and by this assumption vanishes entirely since the lattice as a whole is neutral.

2. Also, the atom core, being a closed shell, is spherical; as a result we have a spherically symmetric—i.e., entirely separable—wave equation to solve within the cell.

3. For the lowest state of the s band, which has of course $k = 0$ and thus a wave function which is simply periodic, $u_0(r)$, the boundary condition takes on the simple form that the slope $d\Psi/dr\,|_{r = R_{cell}} = 0$. This is because the wave function must approach the cell boundary horizontally, since it must by reflection symmetry be even about the boundary. The lowest *atomic* state is the spherically symmetric state for which $d\Psi/dr\,|_{r=\infty} = 0$. The essential physical basis for the cohesion of the alkali metals and in fact for the so-called "metallic bond" is the fact that the energy of the state for which $d\Psi/dr\,|_{r = R_c} = 0$ is considerably lower than the energy of the state in the free atom. Wigner has called this the "boundary correction" and, in his Seitzschrift article (25), estimated it for a wide variety of metals.

We can see immediately from rather crude considerations that this must be so. The spherical symmetry allows us to separate the wave equation $-\nabla^2\Psi + V(r)\Psi = E\Psi$ into angular and radial equations by assuming $\Psi = [U_L(r)/r]\, Y_{LM}(\theta, \phi)$, and the equation for U_L is

$$\frac{d^2U_L}{dr^2} + \left[E - V(r) - \frac{L(L+1)}{r^2}\right] U_L = 0$$

Here the units are atomic ones, r being measured in Bohr radii and all energies in rydbergs. Outside the core of the alkali atom $V = -2Z/r = -2/r$ in this case of a monovalent atom; inside V is more complicated, and also is not in fact a local operator but includes exchange with the core electrons. Nonetheless we may for qualitative purposes treat it as a pure potential $-2Z(r)/r$ (Figure 8).

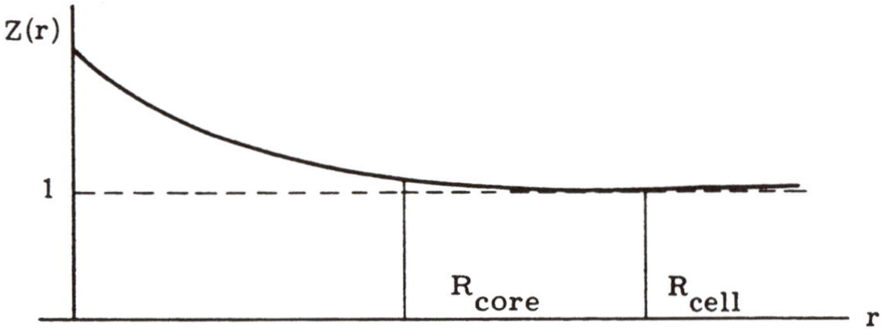

Figure 8

We are interested in the low s states (L = 0), and thus the boundary condition at r = 0 is that Ψ be finite, so that U = 0 and increases linearly. Let us start with some very low energy E_1 and mentally integrate the equation out from this boundary condition at the origin. Before we reach the "classical turning point" at $R_t(E_1)$, $E_1 - V > 0$, and d^2U/dr^2 has the opposite sign from U and the curvature is toward the axis; thereafter, however, $U>0$, $d^2U/dr^2>0$ and the curvature is away from the axis: U blows up. Now let us sketch in a sequence of increasing energies E_1, E_2, E_3, and E_f, the actual free-atom eigenvalue. We see that the free-atom eigenvalue is *above* the energy at which the wave function is flat at R_{cell} and then increases from then on and blows up (Figure 9). The resulting gain in kinetic energy can be rather large. In the case of sodium, for instance, it is 3 1/2 volts; in Li it is higher, in the high alkalies, K, Rb and Cs, lower.

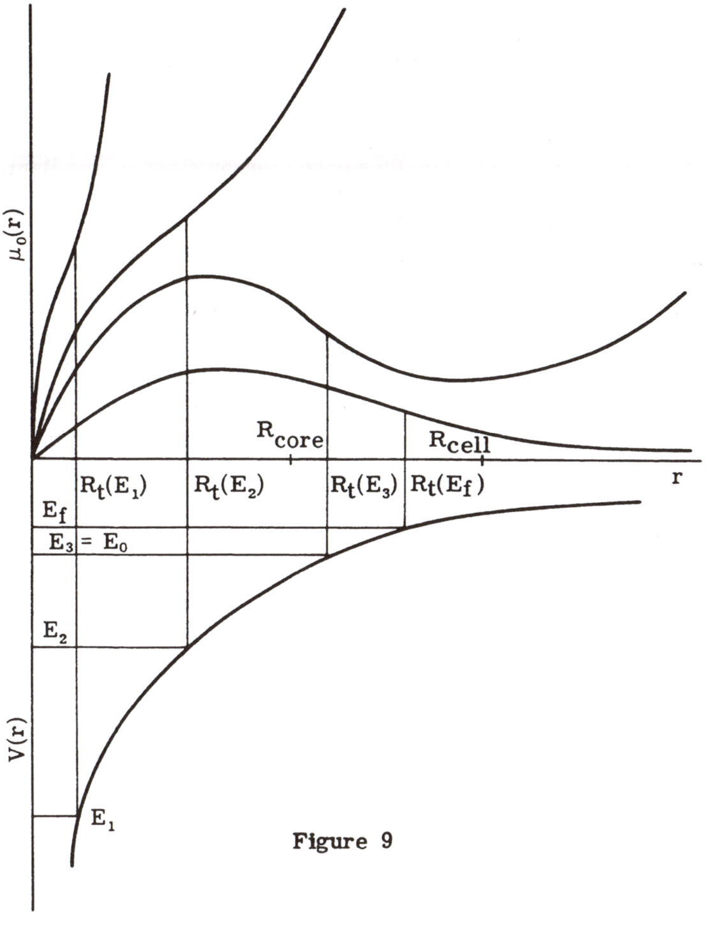

Figure 9

From the diagram it is also possible to get a pretty good idea of the accurate and ingenious method developed by Kuhn, Van Vleck, Brooks (21), and Ham (18) for calculating this energy. On both diagrams I have indicated the fact, valid for all the alkali metals, that the size of the core orbitals (1s in Li, 2s2p in Na, etc.) is very considerably smaller than that of the atomic cell. This means that there is an appreciable region of the radial integration in which we know exactly the potential we must use, namely, $V = -2/r$. Inside the core, of course, the potential is unknown; what was done in the earlier calculations in this region was to start with a Hartree or Hartree-Fock potential and to modify it to fit the observed series of spectral term values, which are known to great accuracy for the alkali atoms.

Kuhn and Van Vleck pointed out that in fact it might never be necessary to integrate a radial equation through this region. If one is to use a set of empirical energy levels from the optical spectrum to adjust one's potential anyhow, might there not be a way to extract the necessary information directly from the optical spectrum itself?

The Kuhn-Van Vleck method is simpler in principle than the more accurate and complete technique which was finally worked out by Brooks and Ham. Their method was to consider the wave function as integrated, starting from $U = 0$ at $R = 0$, out to the point R_{core}. At this point all the information one needs to carry the integration further is the logarithmic derivative $(1/U)\,(dU/dr)\,|_{r = R_{core}}$. This number will be some function of the starting energy E; as we see, it will start out at some high value for large negative E, decrease through zero, and eventually resemble a $\cot^{-1}$ curve. From the point R_{core} on, we know the potential to be of the form $(-\,2/r)$ exactly, and Wannier and others have found analytical forms for the two solutions at an arbitrary energy. Then we can fit these

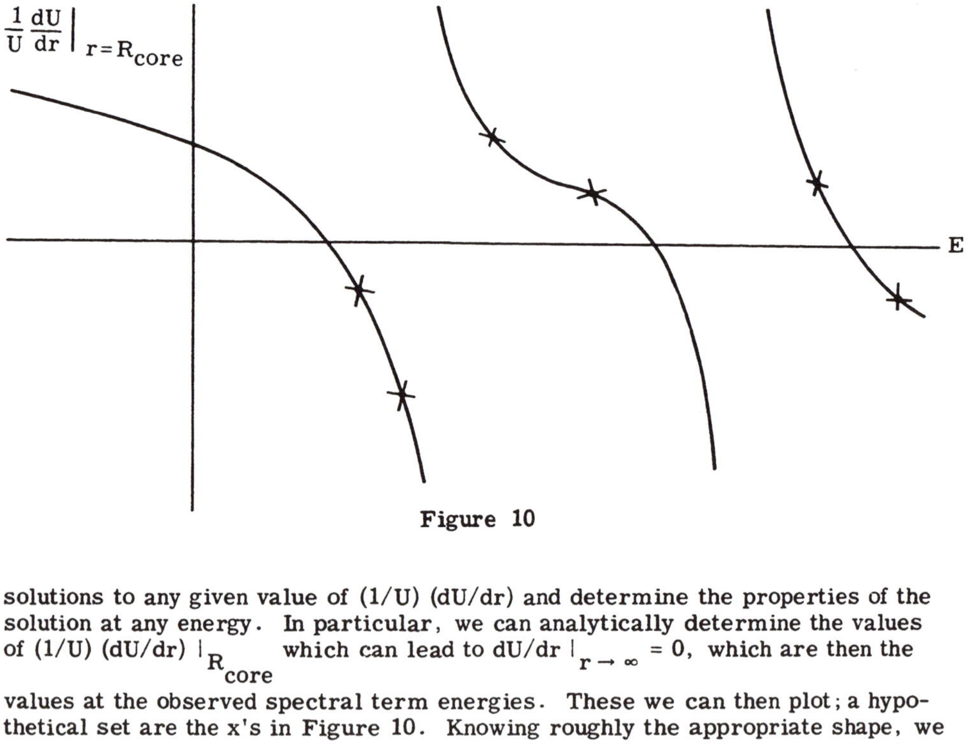

Figure 10

solutions to any given value of $(1/U)\,(dU/dr)$ and determine the properties of the solution at any energy. In particular, we can analytically determine the values of $(1/U)\,(dU/dr)\,|_{R_{core}}$ which can lead to $dU/dr\,|_{r \to \infty} = 0$, which are then the values at the observed spectral term energies. These we can then plot; a hypothetical set are the x's in Figure 10. Knowing roughly the appropriate shape, we can try to draw a smooth curve through these given points, and assume that that is the correct $(1/U)(dU/dr)|_{r=R_c}(E)$.

We also know analytically, again because we know the precise analytic behavior of solutions of the Coulomb wave equation, what value of $(1/U)(dU/dr)|_{R_{core}}$ is necessary to lead to $(1/U)(dU/dr)|_{R_{cell}} = 0$. Call this $(1/U)(dU/dr)|_0$. We then just interpolate our smooth curve to this value, and that gives us a value E_0 of the lowest eigenvalue.

In fact, because of the rather unmanageable shape of the curve for the logarithmic derivative, that is *not* the correct quantity to plot; rather, as Kuhn himself first pointed out, the right quantity to give a really smooth extrapolation is the *quantum defect,* which at the free-atom eigenvalues is just the quantity δ_m in the Ritz interpolation formula for the alkali spectrum,

$$E_m = -\frac{1}{(m - \delta_m)^2}$$

δ_m is closely related to the phase shift δ which occurs in collision theory, but extra complications which are of no interest except to specialists are introduced by the long-range character of the Coulomb potential. In any case, it is found—and can be proved mathematically—that one can find a quantity δ which interpolates smoothly between the observed spectral series terms and which can be used to satisfy any desired boundary condition at the cell surface. In Ham's article are given, to three or four significant figures, the resulting $k = 0$ energies for a number of densities in each of the five alkali metals.

The major advantage of this method, and possibly one of the most important reasons why most cohesive energy—as opposed to actual band shape—calculations by other methods have been unsuccessful, is that the many-body effects in the interaction of valence with core electrons are contained essentially exactly in the empirical values of the quantum defect which are used. It may even be that within R_{core} there is no reasonable local potential $V(r)$ which would give the observed eigenvalue spectrum; as we mentioned, even in H-F the correct potential is nonlocal. This seems in fact to be the main reason why the early Wigner-Seitz calculations were good for Li and Na, but rather bad for K and unworkable for Rb and Cs, while the Q.D.M. seems to have no difficulty with the heavy metals, in which core polarization and exchange are large effects because the cores are larger and less tightly bound.

The location of the one energy level at $k = 0$, then, is done superlatively well by the Q.D.-cellular method; and it is precisely its advantage over all other methods, that at least one level can be located in absolute value accurately. Nonetheless, that cannot be the whole battle; we must first insert the so-called "Fermi energy," the extra kinetic energy which appears because the rest of the electrons have to go into states higher up in the band. Then we must make a set of rather large corrections for electron interactions, the sum of which, however, turns out to be fairly small.

That the Fermi energy is a rather large effect is clear from Wigner's qualitative argument (20), which suggests that a substance with a full s band would be almost unbound. From the point of view of integrating the radial equation, a state at the top of the band must be one which changes sign from the atom to its neighbors, which means that $\Psi = 0$ at R_{cell}, not $d\Psi/dr = 0$ at R_c. The energy of the wave function for which $\Psi = 0$ is probably about as much above the energy of the free-atom wave function, E_I, as E_o is below: One can see that in a sense $\Psi \to 0$ at ∞ is roughly an intermediate criterion between $\Psi|_{R_c} = 0$ and $d\Psi/dR|_{R_c} = 0$ (Figure 11). Thus for a half-filled band we might expect to lose in Fermi energy half or more of the 3 to 4 volts of energy gained in the k = 0 state, and it indeed turns out that we do.

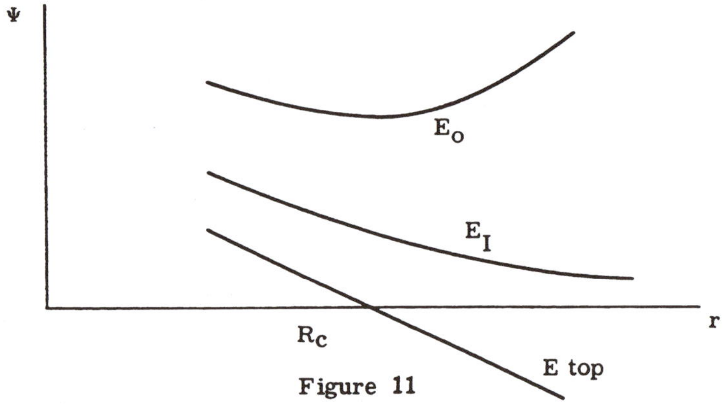

Figure 11

To calculate the Fermi energy it is necessary to have a reasonably accurate estimate of the band structure throughout the occupied region of k space. It turns out that this is a difficult and often inaccurate calculation, using the cellular method, for points near the boundary of the first zone, or for higher bands in which the wave functions no longer are s like in the cells. The reason is that the solutions of the wave equation within the cells are the s, p, d, etc., eigenfunctions of angular momentum with L = 0, 1, 2, etc.... The accuracy of one's calculation or extrapolation decreases rapidly as L increases, so that the fitting of prescribed boundary conditions for a wave function with $L \gtrsim 3$ or 4 is a very difficult matter. This is because of the effective repulsive potential energy $L(L+1)/r^2$ which keeps the electrons out of the core region in the appropriate optical energy levels, while the wave functions needed for boundary-value fitting in the solid problem do penetrate to the core.

It turns out, unfortunately, that band wave functions for reasonably large k contain very appreciable components which must be described by wave functions of high angular-momentum values. The easiest way to see this is Shockley's "empty lattice" test—one may expand symmetrized plane waves, which are a possible set of band wave functions, in spherical harmonics about the center of the cell, and one finds indeed that, for k's near or beyond the zone boundary, they have rather appreciable $L \geq 3$ or 4 components. Although progress has been made in the mathematical problem of fitting spherical harmonics to correct boundary conditions, the accuracy remains questionable for all but the alkali metals.

For the alkali metals, of course, the problem is not very serious, because a large fraction of the electrons are in k values well within the first zone. The trick which is used here is to expand the energy E(k) as a power series in k about the origin; by symmetry the first term is zero and the second, the so-called effective mass term,

$$E = E_o + \frac{h^2 k^2}{2m^*} + \cdots$$

may be obtained by a calculation using the "$k \cdot p$" perturbation theory. The principle of this method is to write down the wave equation satisfied by, not the total wave function $\Psi_k(r)$, but only the periodic part u_k. We start with the usual wave equation

$$\left[-\frac{\hbar^2}{2m} \nabla^2 + V(r) \right] u_k(r) e^{ik \cdot r} = E_k u_k e^{ik \cdot r}$$

Clearly everything but ∇^2 commutes with $e^{ik \cdot r}$, so that the exponential may be factored out after the following computation:

$$\nabla^2 \left(u_k e^{ik \cdot r} \right) = \nabla \cdot \left(\nabla u_k + ik u_k e^{ik \cdot r} \right)$$

$$= -k^2 u_k e^{ik \cdot r} + 2ik e^{ik \cdot r} \cdot \nabla u_k + \left(\nabla^2 u_k \right) e^{ik \cdot r}$$

Thus,

$$-\frac{\hbar^2}{2m} \left(\nabla^2 + 2ik \cdot \nabla \right) u_k + V(r) u_k = \left(E_k - \frac{\hbar^2 k^2}{2m} \right) u_k$$

When we compare this with the equation for the eigenfunction $\Psi_o = u_o$,

$$-\frac{\hbar^2}{2m}\nabla^2 u_o + V(r)u_o = E_o u_o$$

we see that, aside from the term

$$-i\frac{\hbar^2}{m}k\cdot\nabla u_k = \frac{\hbar}{m}k\cdot p u_k$$

it is the same equation and, since u_k is periodic, has the same boundary conditions. Thus we can, for sufficiently small k, treat this term as a perturbation to get an expression for the energy at k in terms of solutions at $k = 0$. In fact, one may use this expression with only slight modifications at any general point k_o to find the wave functions and eigenvalues at neighboring points.

In our particular case of the alkalies, the term $k\cdot p$ contains the gradient operator and so makes an odd function out of an even one and vice versa. This means that the diagonal matrix element $(u_{0n}, (k\cdot p)\, u_{0n})$ vanishes, so that the first-order correction vanishes. The second-order correction is, by standard perturbation theory,

$$E_k - \left(\frac{\hbar^2k^2}{2m} + E_o\right) = -\frac{\hbar^2k^2}{m^2}\sum_n \frac{|(0n|p_x|00)|^2}{E_n(0) - E_o}$$

so that

$$\frac{1}{m^*} = \frac{1}{m}\left[1 - \frac{2}{m}\sum_n \frac{|(0n|p_x|00)|^2}{E_n(0) - E_o}\right]$$

It would appear at first sight that the effective mass always increases—the band becoming flatter—since the sum on the right would be expected to be >0, with $E_n > E_o$. This is indeed true for Li, in which there are no p bands (since p_x is a vector, all the important matrix elements are to p bands) lower than the valence band; but it need not be so in general, and is not in fact in any other alkali except possibly Na. The matrix elements to bands representing closed shells are thus quite important.

Although the original calculation of Wigner and Seitz used the $k\cdot p$ theory directly, Bardeen (27) very shortly introduced a much simpler and more accurate procedure by observing that the first-order correction to the wave function could be calculated essentially exactly. This is done by directly iterating the wave equation using

$$u_k = u_o(r) + u_1^k(r) + \cdots$$

$$\left(-\frac{\hbar^2}{2m}\nabla^2 + V - E_o\right) u_1 = \frac{i\hbar^2}{m} k\cdot\nabla u_o$$

A particular solution for u_1, he observed, is

$$u_1^{(a)} = -ik\cdot r\ u_o(r)$$

(as one finds by substitution; the inhomogenous term and the $\nabla(k\cdot r)\cdot\nabla u_o$ term cancel). That does not satisfy the boundary condition, which is obviously (since u_1 is to have the symmetry of a p function and be periodic) that it vanish on the cell boundary.

We can make the solution vanish on the cell boundary by adding to it a p-function solution of the homogeneous equation,

$$V_1 = \frac{ik\cdot r}{r^2} P(r)$$

where P is a solution at fixed energy E_o of the radial wave equation

$$\frac{d^2P}{dr^2} - \frac{2}{r^2} P(r) + \frac{2m}{\hbar^2}(E_o - V) P = 0$$

Then

$$u_1 = ik\cdot r\left(P/r^2 - u_o(r)\right)$$

Thus

$$\left.\frac{P}{r^2}\right|_{R_c} = \left. u_o\right|_{R_c}$$

We immediately recognise that the quantum-defect method is set up to give us directly, via an interpolation from the optical spectrum of p states, all the interesting properties of the wave function P(r), at least in the outer part of the cell. By using a number of clever transformations, Bardeen was able to write the second-order energy corrections entirely in terms of this first-order correction to the wave function, and even in terms of its behavior at the boundary of the cell:

$$\alpha = \frac{m}{m^*} = \gamma \left[\frac{r}{P} \frac{dP}{dr} - 1 \right]_{R_c}$$

and

$$\gamma = -\frac{1}{3} R_c \left[u_o(R_c) \right]^2 \Big/ \left[\gamma u_o(r) \frac{d}{dr} \left(\frac{1}{r} \frac{\partial u_o}{\partial E} \right) \right]_{R_o, E_o}$$

The proof (partly due to Silverman and others) is rather lengthy, so I shall not discuss it here.

In all the alkali metals the perturbation theory converges quite well, because u_o is quite smooth (the perturbation is $\alpha\,(k \cdot \nabla) u_o$). In Na u_o is practically constant over 90 per cent of the unit cell. Thus the kinetic or Fermi correction E_F is just that for a free-electron gas, corrected for the modified effective mass. This is

$$\int_o^{k_f} k^2 dk \, \frac{h^2 k^2}{2m^*} \Big/ \int_o^{k_f} k^2 \, dk = \frac{3}{10} \frac{h^2 k_F^2}{m^*}$$

where k_F is determined by

$$n = \frac{k_F^3}{3\pi^2}$$

This may be written

$$E_F = \frac{2.21}{r_s^2} \left(\frac{m}{m^*} \right) \quad \text{ryd}$$

where

$$\frac{4\pi}{3} a_o^3 r_s^3 = \frac{1}{n}$$

It will give us a feeling for the orders of magnitude of the various effects to give the various relevant quantities for a number of alkali metals (see Table 3).

Table 3

Element	r_s	$-E_o$	Cohesive boundary correction $E_I - E_o$, ev	$\frac{m}{m^*}$	Kinetic correction E_F, ev	Uncorrected cohesive energy $E_I - E_o - E_F$, ev	Observed cohesive energy, ev
Li	3.21	0.69 ryd = 9.4 ev	4.0	0.67	1.90	2.10	1.65
Na	3.96	0.625 ryd = 8.5 ev	3.15	1.02	2.0	1.15	1.20
Rb	5.20	6.27 ev	1.65	1.10	1.23	0.42	0.85

This is adequate to show the general run of the results. As we guessed, the Fermi kinetic-energy correction tends to run one-half to two-thirds the boundary correction, and the difference accounts in an order-of-magnitude fashion for the binding energy of the metal. Note that the cohesive energy is the smallest energy in the problem, being calculated as the difference of much larger numbers —this is almost always the case and represents the great source of difficulty in binding-energy calculations. It is not hard to calculate bands with an accuracy of 0.1 ryd or so, and that is usually satisfactory for an idea of the general band structure, which is often on the scale of rydbergs, or at least 0.4 to 0.5 ryd; but binding energies are 0.1 or less.

3. Exchange and Correlation in the Free-Electron Gas

At this point you may be a bit surprised to see that we are so close to having calculated the actual binding energy. You may well say: What about the Coulomb interactions of the metallic electrons themselves? We have ostensibly ignored these entirely, in that we have used only the ion core field in calculating our wave functions.

In point of fact, the valence-electron interactions have been taken into account in great part, in a very clever and unobtrusive way. What Wigner has done is in effect to include the interaction of valence electrons when they are on different atoms, because he has made the nearby cells neutral rather than charged +1, as they would be if *only* the ion cores were present; but he has intentionally left out the interaction of electrons in the same cell. This is a very reasonable approximation for the exchange and correlation effects. First, as to exchange—we know that the exclusion principle makes a "hole" in the distribution of parallel spin electrons around a given electron. The Coulomb repulsion tends to force the antiparallel spin electrons as well away from the neighborhood of a given electron, so the net effect is to remove approximately a full electronic charge from the neighborhood of the given charge. The radius of this hole is of the order r_s, so it is reasonable to calculate the wave functions as though only one electron at a time were allowed within each unit cell. The result is that, without solving any very complicated wave equations, the wave functions with which we deal have been calculated to an accuracy practically equivalent to Hartree-Fock. The fact that the hole is relatively smaller in polyvalent metals and that therefore this approximation would not work may be another reason for the difficulty of binding-energy calculations in such metals.

Nevertheless, in order to obtain truly accurate final answers, it is essential to go back, using these wave functions, and attempt a more complete estimate of the "many-body" corrections. We expect the sum of them all to be close to zero for the above reasons, but the individual terms will be fairly large.

What enabled Wigner and Seitz (28, 29) to make a good estimate of these terms is the fortunate fact that the electrons in the alkalies, except for the constant "boundary correction" which lowers the total energy as a whole, behave much like a gas of truly free electrons. As we remarked, in over 90 per cent of the volume of the unit cell the wave functions are practically $e^{ik \cdot r}$ in Na. That allows us to use the theory of the self-interaction energy of the free-electron gas. This theory is greatly simplified by the fact that the gas must be "homogeneous" in space—that is, its properties at each point in space must be the same.

The first correction one applies is the Hartree one, the mean field of all the electrons. In the free-electron case, this is the self-interaction energy of the entire uniform charge distribution, and diverges as $V^{2/3}$; but of course this divergence must be canceled by an equal and opposite term from the ion charges, which ensure neutrality. The usual procedure in studying the free-electron gas is to ignore this term completely; one replaces the ion cores by a uniform "jelly" of positive charge which precisely cancels the mean charge of the free electrons.

In the real alkali metal, on the other hand, there is such a term, large in fact relative to the cohesive energy. We have included the interaction with everything except the central sphere of charge, so we have to include the Coulomb self-energy of a sphere of charge e and radius $R_{cell} = r_s a_o$, which is $3e^2/5R_c = 6/5r_s$ ryd. As you see, in Li that is more than 4 volts and destroys all agreement.

To set against this we must calculate the two effects which tend to keep the electrons apart, exchange per se and correlation. The latter energy is defined, rather artificially, as the difference in energy between the best Hartree-Fock solution and the correct one. We shall not calculate the correlation energy here, but we can do the exchange effect, and, because the mathematics and the ideas will be of use to us later, we shall.

The H-F equation including exchange, realizing that the Hartree term in the case of the free-electron gas is a trivial constant, is

$$-\frac{\hbar^2}{2m}\nabla^2 \phi_{k\sigma} - \sum_{k'\,\mathrm{occ}} \int dr' \; \phi^*_{k'\sigma}(r')\,\phi_{k\sigma}(r')\,\frac{1}{|r-r'|}\,\phi_{k'\sigma}(r) = E_k \phi_{k\sigma}(r)$$

Because the system as a whole still has translational symmetry, we can obtain a self-consistent solution, if not necessarily the lowest one, by assuming

$$\phi_k = \frac{1}{\sqrt{V}}\, e^{ik\cdot r}$$

Then the exchange terms become

$$A\phi_k(r) = \int dr'\ A(r,r')\ \phi_k(r')$$

$$A(r,r') = \sum_{k'} \frac{e^2}{|r-r'|}\ \phi^*_{k'}(r')\ \phi_{k'}(r)$$

$$= \frac{e^2}{V} \sum_{k'} \frac{e^{ik'\cdot(r-r')}}{|r-r'|} = A(r-r')$$

This does have translational symmetry (doesn't depend on r + r') and therefore the solutions of the equations can indeed be plane waves.

We can actually compute the form of A(r - r'), which is often called the potential of the exchange hole. Much of this development is due to Dirac (30). We can think of it as the potential of a "hole" in the charge density surrounding the electron at r':

$$\rho(r-r') = \frac{e}{V} \sum_{k\ \mathrm{occ}} e^{ik\cdot(r-r')}$$

$$= \frac{e}{V}\left(\frac{n}{2}V\right) \int_0^{k_F} k^2\ dk \int_{-1}^{1} dx\ \ e^{ik|r-r'|x} \Big/ 2\int_0^{k_F} k^2\ dk$$

Using

$$\frac{n}{2} = \frac{4\pi}{3}\ \frac{k_F^3}{8\pi^3} = \frac{k_F^3}{6\pi^2}$$

and $R = |r-r'|$, this is

$$= -\frac{e^2}{2\pi^2 R}\ \frac{d}{dR}\left\{\int_0^{k_F} \cos kR\ dk\right\} = -\frac{e^2}{2\pi^2 R}\ \frac{d}{dR}\left(\frac{\sin k_F R}{R}\right)$$

$$\left(\text{since } \frac{\sin kR}{R} \rightarrow 0 \text{ as } k \rightarrow 0\right)$$

$$\rho(R) = \frac{ek_F}{2\pi}\left[-\frac{\cos k_F R}{(k_F R)^2} + \frac{\sin k_F R}{(k_F R)^3}\right]$$

A picture of the exchange hole is given in Figure 12. (Note that the singularities at $R = 0$ cancel out.) This plot shows the total effective density of parallel spin electrons at R seen by an electron at zero.

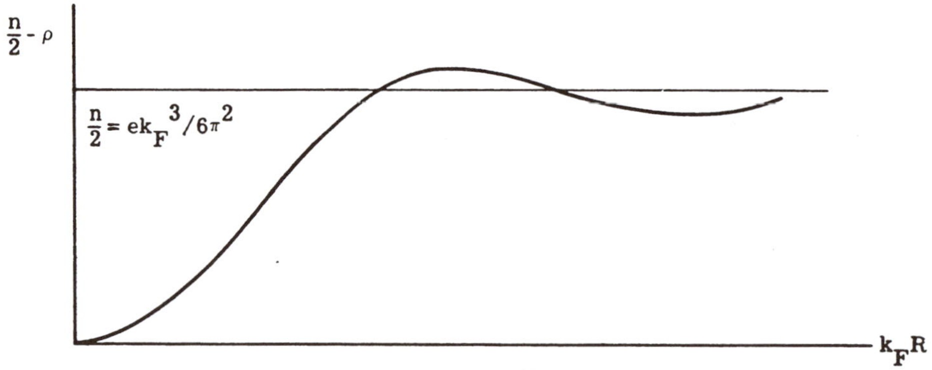

Figure 12

The corresponding kernel, the exchange potential $A(r-r')$, is of course $e\rho$ divided by $R = r-r'$:

$$A(R) = \frac{e^2 k_F{}^4}{2\pi^2}\left(\frac{\sin k_F R}{(k_F R)^4} - \frac{\cos k_F R}{(k_F R)^3}\right)$$

These functions, with their characteristic very long range oscillatory behavior at ∞, are extremely common and very useful in the theory of the free-electron gas. A is similar to the "Green's function" of the free-electron gas, and those of you familiar with spin resonance will recognize in $A(r)$ the Ruderman-Kittel function. The long-range, oscillatory behavior is the result of the sharp upper limit in the integration over the Fermi distribution, and occurs only when there is a free Fermi surface. These two functions contain $2k_F R$ rather than $k_F R$ because they result from a slightly different integral, but the reason for the oscillators is the same.

The exchange contribution to the energy of a given plane-wave state k is easily deduced from

$$A\phi_k(r) = \int A(r-r')\phi_k(r')\, dr'$$

which, from the convolution theorem for folding a Fourier transform, is

$$A\phi_k = A_k \phi_k$$

where

$$A_k = \int d\mathbf{R}\ e^{ik\cdot\mathbf{R}} A(R)$$

This Fourier transform may be calculated directly, or as follows: $A(r) = \rho(r)/r$, so that, again using the convolution theorem,

$$A_k = \frac{e}{(2\pi)^3}\int dk'\ \rho(k)\left(\frac{1}{r}\right)_{k-k'}$$

$$= \frac{4\pi e}{(2\pi)^3}\int dk'\ \rho(k')\Big/|k-k'|^2$$

But $\rho(k')$ is just e/V when k is occupied, $k < k_F$, 0 otherwise; so

$$A_k = \frac{4\pi e^2}{(2\pi)^3 V}\sum_{k'\ \mathrm{occ.}}\frac{1}{|k-k'|^2} = \frac{e^2}{2\pi^2}\int\frac{d^3k'}{|k-k'|^2}$$

This integral, so far as I know, must be tediously evaluated by direct means; it was done by Dirac in 1930. The result is

$$A_k = \frac{0.612}{r_s}\left[2 + \frac{k_F^{\,2} - k^2}{k\,k_F}\ \ln\left|\frac{k+k_F}{k-k_F}\right|\right]$$

This integral is a very famous and useful function. We may sketch the quantity in brackets as follows: Its value at $k = 0$ is 4; the singularity in the denominator is canceled by $\ln|k+k_F/k-k_F| \to 2k/k_F$. At $k = k_F$ there is a second singularity, but it is of the form $0 \ln 0 = 0$ so $A = 2$, but the slope is infinite at that point (Figure 13).

We can think of this curve as a smudged-out replica of the Fermi distribution function, with the infinite slope corresponding to the sharp drop at the Fermi surface. The reason for it is the fact that when $k < k_F$, the denominator goes to zero at $k' = k$; but when $k > k_F$, it does not. The zero is an integrable one but just barely so. That in turn reflects the long-range nature of the Coulomb potential: the Fourier transform of $1/r$ is $2\pi/k^2$.[1]

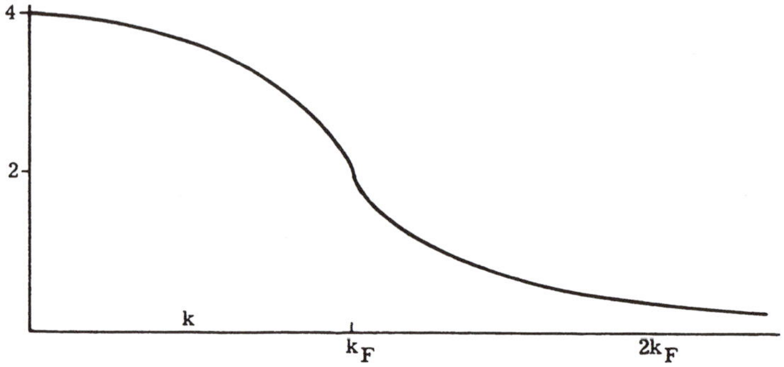

Figure 13

From A(k) we can calculate two things: First, the mean exchange energy, which will be $(1/2N)\ \Sigma_k A(k)$. We see A varies from (2 to 4) times $0.612/r_s$; the average factor is about 3: $E_{exch} = -\ 0.916/r_s$, which cancels out a large proportion of the $1.2/r_s$ Coulomb self-energy. Nonetheless, the difference $0.25/r_s$ ryd $\cong$ 0.85 ev in Na is enough to remove a large fraction of the cohesive energy.

The remaining correction is the "correlation energy," which is the correction which must be made because electrons even of antiparallel spin tend to stay away from each other. There is an enormous literature on this correction term; suffice it to say that Wigner and, later, Pines (31), by interpolating over a rather wide range between expressions which are certainly good for $r_s \lesssim 1$ and others which are equally good for $r_s \gtrsim 10$, but neither of which has any reasonable hope of being

---

[1] The infinite slope can be directly demonstrated by taking

$$\nabla_k A_k = \frac{4\pi e^2}{V\delta k}\left[\sum_{k\ \mathrm{occ.}} \frac{1}{|\mathbf{k}-\mathbf{k}'|^2} - \sum_{(\mathbf{k}+\delta\mathbf{k})\ \mathrm{occ.}} \frac{1}{|\mathbf{k}-\mathbf{k}'|^2}\right]$$

$$= -\frac{4\pi e^2}{V} \sum_{\text{surface of Fermi sphere}} \frac{\cos\theta}{|\mathbf{k}-\mathbf{k}'|^2}$$

The surface element goes like k dk, so we get a logarithmic divergence $\int k\, dk/k^2$.

close for $r_s = 4$, obtain the following estimate. From the fact that it is not rapidly varying in the known range and that various estimates agree, we can guess it to be valid within ± 20 per cent: $E_{corr} = 0.88/(r_s + 7.8)$ ryd. [This is not what has been used by many authors, which is an earlier expression of Wigner's which he corrected in a footnote in 1938 (29).] With this expression we obtain the results given in Table 4.

Table 4

Element	Uncorr. E_{coh}, ev	ΔE_{HF}, ev	ΔE_{corr}, ev	Total, ev	Expt., ev
Li	-2.1	1.19	-1.09	-2.0	-1.65
Na	-1.15	0.96	-1.02	-1.2	-1.2
Rb	-0.42	0.73	-0.92	-0.61	-0.85

I should caution you that the tables I give here are not authoritative and do not represent the most exact or best results which have been achieved by the Q.D.M. For more detailed results I refer you to the actual papers, especially Ham's article and that of Brooks in the Varenna Summer School report. As pointed out in this last paper, the good agreement achieved by Ham is lessened by various further corrections. The only sure conclusions which can be drawn are perhaps these: that Wigner's estimate of the correlation energy for free electrons is not far from the correct value; and that we do indeed know the true physical origins of the cohesion in the alkalies. Corrections which are of the order of the error of ± 0.2 ev in the correlation energy might be further terms in the Fermi energy, and most importantly the corrections to exchange and correlation arising from the fact that the wave functions are not free-electron ones. As we may see later, it is not legitimate just to insert m^* in our formulas for these effects, and the true nature of these corrections is not as yet known. [See, however, Phillips' work, to be discussed shortly (25).] These corrections could easily be of order 20 per cent of the correlation energy.

One aspect of the numerical results of the Q.D.M. which is striking and very encouraging is the almost equally good agreement with that on the energy, which is obtained on the compressibility of these metals. Many earlier calculations, and most calculations by other methods, give very bad compressibilities. After all, the compressibility is a second difference of the binding-energy curve. The reason the Q.D.M. is so good is probably that we do know quite well the form of the dependence on density of all the terms: the E_0 term, because it can be calculated so accurately; the rest, because it is determined by general theoretical considerations. The kinetic energy goes as $1/r_s^2$, the exchange, Coulomb, and correlation energies as $1/r_s$, roughly. Often a very good fit to the pV curves is obtained with Bardeen's semiempirical equation of state based on this idea (27):

$$E = \frac{A}{r_s^3} + \frac{B}{r_s^2} - \frac{C}{r_s}$$

Now I should like to return briefly to the exchange energy before going on to discuss the O.P.W. method. The other aspect of A(k) which is important is that it represents a k-dependent correction to the self-energy E_k of the electron of momentum k: $E_k = (h^2k^2/2m^*) - A(k)$. This is the characteristic behavior of a nonlocal potential such as exchange. The property of being nonlocal is usual for effective potentials caused by the various kinds of many-body effects. The potential is $A\phi = \int A(r - r')\phi(r')\, dr'$; by the convolution theorem, this is transformed into a k-dependent correction $A\phi_k = A_k\phi_k$ to the self-energy, which can be thought of as an *effective mass* correction in a rough kind of way, caused not by the periodic potential of the lattice but by the change in the exchange hole with momentum of the particle.

In the special case of exchange, using the Coulomb interaction, the effective mass correction diverges just at the Fermi surface: Since A(k) has an infinite negative slope, $dE_k/dk|_{E_F} = +\infty$ precisely at the Fermi surface. The density of states falls to zero logarithmically. This would, if correct, be in contradiction with many experimental results on low-temperature specific heat, spin paramagnetism, etc. Wigner pointed out in the original papers that in fact the long-range part of the Coulomb interaction would be screened in a real electron gas by the slight displacements of the other electrons nearby which also are involved in the correlation energy. A rough calculation of this potential screening leads to an effective potential $e^{-k_0 r}/r$, where k_0 is the Debye wave number, of the order of k_F. Then the singularity in A(k) gets smeared out and the effective mass correction is not divergent. We should emphasize that the "exchange hole" is still a long-range, oscillating function with a singularity in k space at the Fermi surface, but the effect of this on the exchange potential A is smeared out by shortening the range of interaction. The curve for A(k) vs. k then looks something like Figure 14.

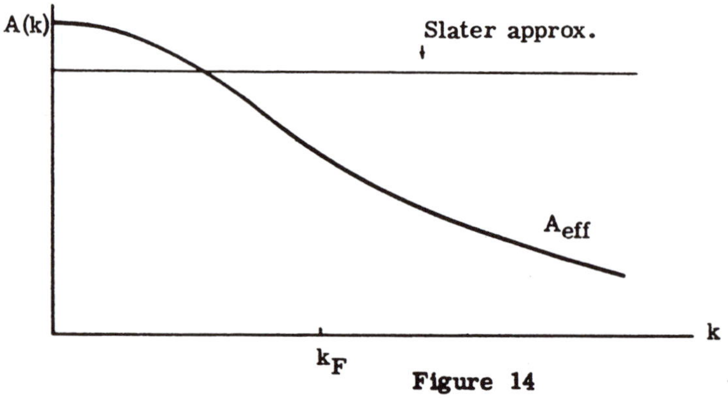

Figure 14

The net result is a decrease in effective mass whose value may be smallish (5 to 20 per cent) at metallic densities.

In a real crystal the density of free electrons is not constant, the free-electron model is not valid, and A is no longer A(r - r') but A(r, r'); it is both nonlocal and space-varying. An approximation which is often made and is due to Slater (32) is to neglect the nonlocal character of A in this case, and to try to replace it by the best local potential $A(r)\,\delta(r-r')$, which will then be proportional to the 1/3 power of the local density. The δ function has a constant Fourier transform, so what is being done is replacing A(k) by an appropriate average (see Figure 14). The criterion for this being a workable approximation is that the density of electrons be slowly varying compared to the size of the exchange hole, which is roughly r_s; then the nonlocal potential will sample regions of roughly constant exchange potential. Such an approximation would be unthinkable in the alkalies, where $r_s > r_{core}$ and thus >> lengths in which variation occurs, but may be better in polyvalent substances where $r_s \sim 1 - 2$. A rather exhaustive study of this question by Phillips in diamond, with $r_s = 1.3$(33), showed that the Slater approximation was not serious there, although there were 20 to 30 per cent corrections to some aspects of the band structure, notably to the over-all height of the Fermi surface from the lowest state k = 0, L = 0. This is, of course, an effect which is always to be expected, corresponding to the momentum dependence of A(k) for the true free-electron gas.[1]

4. The O. P. W. Method

Now we go on to study a method which has as yet never been very successful in calculating binding energies, but which has been remarkably successful in calculating band structures. The fortunate re-emergence of this powerful method at the same time that experimental methods of a number of kinds became available for the detailed investigation of complicated band structures—and all band structures except the alkalies are complicated— has led to a complete revolution in the last 10 years in the understanding of the behavior of electrons in solids.

The O.P.W. method was invented by Herring in 1940 (34) and used by Herring and Hill (35) in their calculation of the binding energy of Be. The fact that the resulting binding energy was not very good rather obscured the fact that the calculation probably gave a very good band structure — as Herring remarks in the original paper. The basic idea is the following: One has very little hope that a calculation based on the nearly free electron perturbation theory — i.e., on true plane waves as starting eigenfunctions — can possibly be practicable in a real solid. This is true for two reasons. The first is that the potential of the atom cores

---

[1] Probably the first comment on nuclear matter as a Fermi gas was made by Van Vleck, in pointing out that in that case also the exchange effective mass correction might lead to a lighter m* — an effect which occurs and is vitally important.

actually becomes extremely large in the region near the nuclei, so that there is no hope that it can be treated as a perturbation. In some sense we must introduce a more nearly exact solution in this region, because the true eigenfunctions must vary rapidly there, and that variation is describable only in terms of plane waves with very high k vectors.

The second reason is even more convincing. It is that even if we were to succeed in doing the calculation using plane waves, to infinitely high order, the result would probably not converge to the right answer for our purposes. That is because in all real metals there are a number of core levels — in Be or C, the $(1s)^2$, in more complex atoms many more, which have very large binding energies relative to the valence electron levels we are after. Thus at any k in the Brillouin zone there are one or more core band levels, to which the lowest eigenvalue of a correct calculation will tend. As is well known, the calculation of higher eigenvalues is a relatively difficult matter.

Herring's solution to this problem was the almost obvious one of taking as starting wave functions not plane waves, but plane waves orthogonalized to the wave functions of the core levels. This will at the very least solve the second difficulty if the core wave functions are reasonably well chosen. It is fortunate that because of the fact that the core levels are mostly in the region of the strong and relatively well-known potential coming from the nucleus, even the Hartree approximation is really quite accurate for them, since electron-electron interaction effects are not very serious. Thus the choice of core functions is not a sensitive problem.

It was already noticed by Herring that the choice of orthogonalized plane waves as starting functions led to at least a partial solution of the first problem, also. In orthogonalizing the smoothly varying plane-wave functions to the core levels, one introduces into the plane waves very rapid variations peaked up near the nucleus. It was at first merely observed empirically that this rapid variation seemed to be just such as to make the starting function a passable solution of the wave equation in the strong region of the core potential, so that the procedure did seem to converge at a manageably rapid rate.

Let us now write down the theory formally. The core functions obtained in some way such as from a Hartree-Fock theory for the free atoms we shall call

$$X_n (r - R_j)$$

where this is the nth core orbital about the jth atom. If X is really a core orbital, it must be an adequate approximation that the overlap from R_j to $R_j + a$ is negligible. (Herring devised ways of treating any residual overlap perturbation theoretically.) Thus

$$\left(X_n (r - R_j), X_m (r - R_k)\right) = \delta_{nm} \delta_{jk} \tag{13}$$

Then the corresponding band functions for a given k within the first Brillouin zone are just the "tight-binding" functions

$$\phi_k^n = \frac{1}{N^{1/2}} \sum_j \exp(ik\cdot R_j) X_n(r - R_j)$$

These too are orthogonal and normalized, if (13) is satisfied.

The various possible plane waves which belong to a given k in the first BZ may be labeled by the reciprocal lattice vectors K:

$$\phi_{k,K}(r) = \frac{1}{V^{1/2}} e^{i(k+K)\cdot r}$$

The orthogonalized plane wave function is then

$$\Psi_{k,K}(r) = \phi_{k,K}(r) - \sum_n (\phi_{k,K}, \phi_k^n)\, \phi_k^n(r) \tag{14}$$

where

$$(\phi_{k,K}, \phi_k^n) = a_{K,n} \tag{15}$$

is the overlap integral between the plane wave and the core function:

$$a_{K,n} = \frac{1}{\sqrt{NV}} \int dr \ \exp\,[-i(k+K)\cdot r] \sum_j e^{ik\cdot R_j} X_n(r - R_j)$$

$$= \frac{1}{\Omega^{1/2}} \int dr \ e^{-i(k+K)\cdot(r - R_j)} X_n(r - R_j)$$

$$= \frac{1}{\Omega^{1/2}} \int dr \ e^{-i(k+K)\cdot r} X_n(r)$$

This is just the $(k + K)^{th}$ Fourier component of X, or, if you like, it is the amplitude of the momentum-space wave function $X_n(p)$ at $p = k + K$:

$$a_{K,n} = X_n(p = k + K)$$

Unfortunately, the orthogonalized plane waves, Eq. (14), are neither normalized nor orthogonal among themselves. Thus we must use the standard treatment for nonorthonormal functions: We must solve for the eigenvalues E of the secular determinant

$$\left|(\Psi_{k,K'}\ (\mathcal{H} - E)\ \Psi_{k,K'})\right| = 0 \tag{16}$$

in which E will appear in some off-diagonal elements also. The general element of (16) may be computed, at the cost of considerable labor, and is

$$E\,(\delta_{KK'} - \sum_n a^*_{Kn} a_{K'n}) = E^o_{k+K}\,\delta_{KK'}$$

$$- (E^o_{k+K} + E^o_{k+K'}) \sum_n a^*_{Kn}\, a_{K'n} + V_{KK'}$$

$$- \sum_n \left\{ a^*_{Kn}\,(X_n,\ V\phi_{k,K'}) + a_{K'n}\,(\phi_{k,K'}\ VX_n) \right\}$$

$$+ \sum_n a^*_{Kn}\, a_{K'n}\, E_n$$

E_n is the nth core-energy level. We have assumed that the core bands are very narrow, which is implied by the vanishing overlap. E^o_k is the free-electron kinetic energy $\hbar^2 k^2/2m$. As you can see, this is a rather complicated expression, even though each of its terms is in principle directly calculable, once we know V(r) and the core levels.

The procedure is to solve, of course, not the infinite determinant (16) but a finite one obtained by judiciously truncating the number of plane waves which we consider. In general, one may still be left with a rather large secular determinant to solve — although with modern machines quite large numbers of plane waves may be used, up to 20 or more — but in order to get results easily and to test convergence it is usual to concentrate on a few special symmetry points of the Brillouin zone. At a symmetry point, we can form symmetrized combinations of plane waves, as we discussed earlier in connection with the almost-free electron model, of different symmetry types; and the symmetry types cannot be mixed by the operation of the periodic potential.

You will remember, for instance, that for the f.c.c. space lattice, and k = 0, we could form the following sets of symmetrized plane waves, all coming from linear combinations of the eight functions corresponding to $K = 2\pi/a\ (\pm 1, \pm 1, \pm 1)$:

$$\Gamma_1 \ : \ \cos\frac{2\pi x}{a} \quad \cos\frac{2\pi y}{a} \quad \cos\frac{2\pi z}{a} \qquad 1$$

$$\Gamma_2' \ : \ \sin\frac{2\pi x}{a} \quad \sin\frac{2\pi y}{a} \quad \sin\frac{2\pi z}{a} \qquad 1$$

$$\Gamma_{15} \ : \ \cos\frac{2\pi x}{a} \quad \cos\frac{2\pi y}{a} \quad \sin\frac{2\pi z}{a} + \text{permutations} \qquad 3$$

$$\Gamma_{25}' \ : \ \cos\frac{2\pi x}{a} \quad \sin\frac{2\pi y}{a} \quad \sin\frac{2\pi z}{a} + \text{permutations} \qquad 3$$

Of all these, only Γ_1 connects with the K = 0 plane wave $\phi_0 = 1$. Thus by solving a 2 X 2 equation for Γ_1, and no further equations except the calculation of diagonal elements, we have taken into account nine plane waves. We could add to this the six plane waves of $K = 2\pi/a\ (\pm 2, 0, 0)$ etc., by including six symmetrized plane waves

$$\Gamma_{15} \ : \ \sin\frac{4\pi x}{a}, \ \text{etc.} \qquad 3$$

$$\Gamma_1 \ : \ \cos\frac{4\pi x}{a} + \cos\frac{4\pi y}{a} + \cos\frac{4\pi z}{a} \qquad 1$$

and a two-dimensional representation Γ_{12}: $\cos(4\pi x/a) - \cos(4\pi y/a)$ (2). Then we would have included 15 plane waves effectively with no more than third-order secular equations, and that only in the case of the three Γ_1 waves.

Unfortunately, together with this advantage we have the disadvantage that at such a symmetry point the orthogonalization process is not only simplified, but oversimplified. In the above example, imagine that the substance under study had only 1s core electrons (this is a fictitious case but adequate to show the principle). Then the only sets of orthogonalized plane waves which are not automatically orthogonal to the cores by symmetry are those of symmetry Γ_1: $\cos(2\pi x/a) \times \cos \times \cos$; $\cos(4\pi x/a) + \cos + \cos$; and the plane wave k = 0. These must all be orthogonalized to the 1s functions, and the full secular determinant worked out, although now, of course, it is not very complicated. In all other cases, we work directly with plane waves alone, for which the secular determinant is merely

$$\left| (E_{k+K} - E)\,\delta_{KK'} + V_{KK'} \right| = 0$$

and is at most of second order.

In general, this will happen at any symmetry point for some sets of plane waves. In Si, there are 1s, 2s, 2p orthogonalizations to be made, the former two to the Γ_1 sets, the latter to Γ_{15}. On the other hand, the sets Γ_{25}' and Γ_{12} are still automatically orthogonal to all cores, because they have d-like symmetry near the atom; and Γ_2 is like xyz, a state of f-like symmetry: no orthogonalizing need be done short of Lu for this.

While saving a great deal of trouble, this also causes trouble because these functions then retain their precise plane-wave character, without any atomic-like appearance near the cores of the atoms. In order to give them the rapid variation near the core which is characteristic of Bloch functions, a relatively large number of plane waves must be introduced before truncation; the convergence of the O.P.W. method for these functions is relatively slow. It is not quite as slow as one might fear, because already from this example we see that we must go to rather large K vectors to find waves with which these combine; there is a physical reason for this which we shall understand shortly.

It is, from the start, clear physically that there must be a great deal of cancellation of large terms in the secular equation (16). The matrix elements $V_{KK'}$ are such that without the orthogonalization terms the resulting energy would be the very large binding energy of the core levels — in diamond, -22.7 ryd, for instance. In some way the orthogonalization terms succeed in assuring that no wave function is bound with an energy greater that - 2 ryd or so; this is the lowest energy of an atomic function in the valence shell. Thus in some sense all but 10 per cent of the enormous attractive potential energy V(r) has been canceled out, for functions which are forced to be orthogonal to the core levels.

The underlying theory of this cancellation, which I shall call "Phillips' cancellation theory," has been sketched by Phillips and presented very nicely by Cohen and Heine (36). I shall follow the latter discussion here. A further important reference is Austin, Heine, and Sham (37).

The basic idea of cancellation theory is that we should try to find the wave equation which is satisfied, not by the true wave function Ψ_k, which is $[(p^2/2m) + V(r)]\ \Psi_k = E_k\Psi_k$, but by what Phillips calls the "smooth" part of Ψ— namely, in this case the plane-wave part ϕ, from which the orthogonalization terms are to be subtracted.

Let us then do this. If the exact wave function is Eq. (14):

$$\Psi_k = \phi_k - \sum_n \left(\phi_k, \phi_k^n\right) \phi_k^n$$

(ϕ_k^n being the core levels), we have

$$\left(\frac{p^2}{2m} + V\right)\Psi_k = \left(\frac{p^2}{2m} + V\right)\phi_k$$

$$- \sum_n \left(\phi_k, \phi_k^n\right)\left(\frac{p^2}{2m} + V\right)\phi_k^n = E_k\left(\phi_k - \sum_n \left(\phi_k, \phi_k^n\right) \phi_k^n\right)$$

or,

$$\left(\frac{p^2}{2m} + V + V_R\right)\phi_k = E_k \phi_k$$

where

$$V_R = \sum_n (E_k - E_n) \frac{\phi_k^n}{\phi_k} (\phi_k, \phi_k^n)$$

V_R, thus defined rather arbitrarily by dividing through by ϕ_k, takes on a real meaning when we see that ϕ_k varies, we hope, very little in the core region relative to ϕ_k^n, so that the division by ϕ_k just gives a constant times ϕ_k^n; and that E_k is small in general relative to a core binding energy E_n, so that its variation is negligible. Thus V is something like a true potential.

What it really is is made clear by Cohen and Heine: It is our old friend, a nonlocal potential. Namely,

$$V_R \phi_k(r) = \int dr' \; V_R(r, r') \, \phi_k(r')$$

$$V_R(r, r') = \sum_n (E_k - E_n) \, \phi_k^n(r) \, (\phi_k^n(r'))^*$$

as we see immediately by integration. The contention of Phillips, essentially by the above physical reasoning, was that V_R was a repulsive potential (since E_n is negative) which often very nearly canceled V.

Cohen and Heine demonstrated this cancellation mathematically in the following way: Looking at the equation (14), which in essence defines the "smooth" wave function ϕ_k in terms of the exact wave function Ψ_k, we see that this definition is by no means unique. After all, in the application we are thinking of, the set of plane waves ϕ_{k+K} is a complete set, so that even the functions ϕ_k^n can be expanded in terms of them. Therefore, we could actually write Ψ_k entirely in terms of the set of plane waves without orthogonalization if we liked, but that would of course give us a very nonconvergent plane-wave expansion; or we could in principle divide the orthogonalization terms in any proportion we liked between ϕ and the explicit terms.

This means that we must define what we mean by "smooth wave function φ" by some extra, explicit condition. We know what we mean by it actually: It's the nearest thing to a few plane waves we can get away with. Cohen and Heine point out that a mathematical way to express that idea is to demand that the effective potential $(V + V_R)$ be as weak as possible:

$$(\varphi,\ (V + V_R)\ \varphi\ )/(\varphi, \varphi) = \text{minimum}$$

From this condition, by straightforward mathematics, one obtains the following equation for $(V + V_R)\varphi$:

$$(V + V_R)\varphi = V\varphi - \sum_n (\varphi_k^n,\ V\varphi)\ \varphi_k^n$$

$$+ (\varphi,\ [V + V_R]\ \varphi\ ) \sum_n (\varphi_k^n,\ \varphi)\ \varphi_k^n$$

so that the diagonal element of this effective potential is

$$(\varphi,\ [V + V_R]\ \varphi\ ) = \frac{(\varphi,\ V\varphi)\ -\ \sum_n (\varphi, \varphi_k^n)\ (\varphi_k^n,\ V\varphi)}{1 - \sum_n \left|\ (\varphi_k^n,\ \varphi)\ \right|^2}$$

The corresponding nonlocal potential is

$$(V + V_R)\ (r, r') = V(r) \left\{ \frac{\delta\ (r - r') - \sum_n \varphi_k^{*n}\ (r)\ \ \varphi_k^n\ (r')}{1 - \sum_n \left|(\varphi,\ \varphi_k^n)\ \right|^2} \right\} \tag{17}$$

(*Note:* Austin, Heine, and Sham (37) have shown since that the denominator is unnecessary.)

Thus we have proved two things: that the effective potential we are using is the weakest possible in the above sense; and that it can be expressed as in Eq. (17). Now note that

$$\sum_{\text{all states}} \varphi_k^{*n}\ (r)\ \varphi_k^n(r') = \delta(r - r') \tag{18}$$

Thus except for the normalization factor in the denominator, the potential *is as effectively canceled as it can be by a linear combination of bound states.* Another way of putting it is that the cancellation is proportional to the degree to which the bound states form a complete set in the region where V is large. Cohen and Heine give a remarkable diagram for Si which shows how good the cancellation can be. (Figure 15) (In drawing the diagram as a local potential advantage is taken of the fact that for s states ϕ is, for purposes of integration with a core function, a constant, so that the nonlocality *for a given* l is not important).

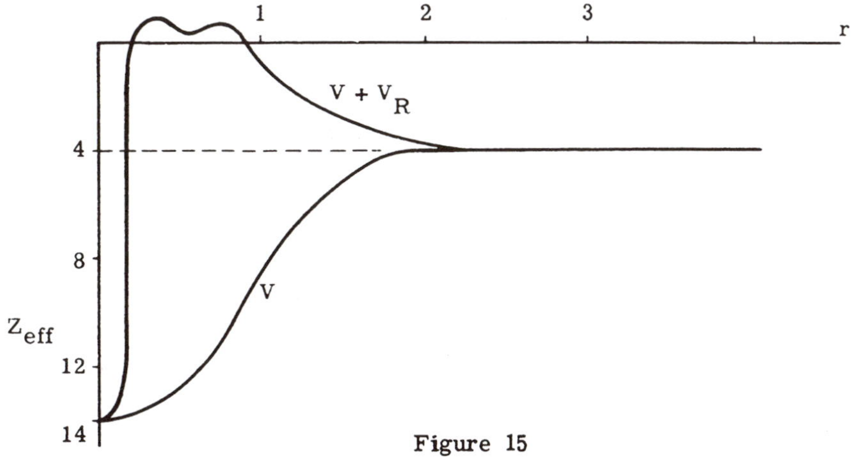

Figure 15

Typically, the normalization factor $1/1 - \sum_n \left|(\phi, \phi_k^n)\right|^2$ is rather close to 1: about 1.1 or so (see note above).

Notice that the nonlocality of the potential is very relevant in making the potential behave quite differently for different l values. For instance, as we already pointed out, there is no cancellation for p levels when none of the core levels are p-like; we can see this by observing that in this case

$$\int dr' \, V_R(r - r') \phi_p (r') = 0$$

since this is the same as

$$\int dr' \, V(r) \phi_{1s}(r) \phi_{1s}(r') \phi_p(r')$$

Here ϕ_p is meant to indicate some symmetrized plane wave which necessarily has p symmetry, such as a Γ_{15} combination. In general, the cancellation is less complete for higher and higher l values. There is a very obvious reason for this when we look at the radial wave equation for an orbital with angular momentum L:

$$\frac{d^2u_L}{dr^2} - \left[\frac{(L(L+1))}{r^2} + V(r)\right] u_L(r) = E_L u_L(r)$$

and sketch the potential-energy term (Figure 16). We see that the centrifugal potential term excludes bound states from the interior region of V, so that for larger L bound states cannot possibly represent a complete set for the very region where V is strongest. Nonetheless, although this idea does not appear explicitly in the

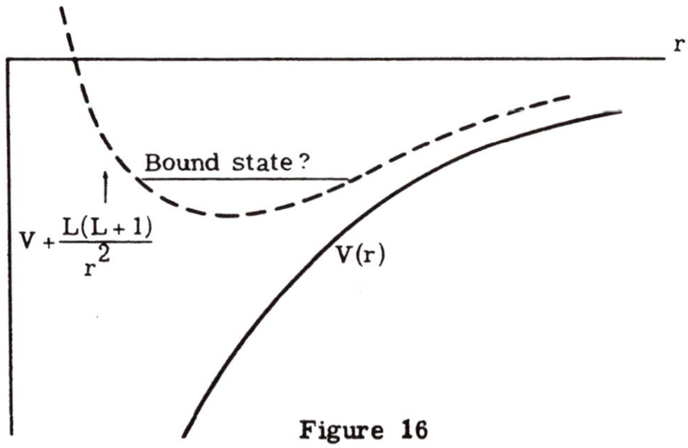

Figure 16

mathematics, the large matrix elements of V cannot really be very effective for these higher L states because the plane-wave states do not penetrate the core very effectively either—the centrifugal term acts also for them. It appears, then, that some of the cancellation which must occur comes from rather high k plane waves, which is part of the reason for the poor convergence in these states. Actually, Heine has remarked that a certain degree of cancellation appears in the secular equation explicitly.

I believe that there remains here the possibility for further improvement of this cancellation theory. The nuclear physicists among the readers may have noticed the remarkable similarity of some of these ideas—both in the Q.D.M. and in the "cancelled O.P.W."—to the techniques of multiple scattering theory and the effective range theory and the use of "T", the scattering matrix, as an effective potential in that theory (38). I think that probably more can be done in analogy with the multiple scattering theory, especially in eliminating high-momentum continuum states as well as the bound states from the problem.[1]

---

[1]Since writing these notes it has come to my attention that the Green's function method of Ham and Segall [*Phys. Rev.*, **124**. 1786 (1961)] is a first step in this direction: probably the excellent convergence of this and the augmented plane-wave method (M. M. Saffren, Thesis, N.I.T., 1959) are due to their similarity to multiple-scattering theory.

The detailed quantitative work which has been done using O.P.W. with cancellation is a subject for the specialist. I should, on the other hand, like to call attention to some of the qualitative features of the behavior of solids which it helps to explain. There are essentially two categories of these: (1) aspects of the behavior of "good metals" which are explained by it; (2) aspects of the quantum chemistry of solids which are explained by when O.P.W. is good and when it fails.

1. The more familiar to Cambridge people of these aspects is the remarkable success of the nearly free electron picture in qualitatively explaining the observed Fermi surfaces of pure metals. I think this aspect is well enough known to need no further discussion. Of the conspicuous failures of this generalization, copper is a case in which the relevant core levels—the d levels—are not far enough down to be treated by orthogonalization, and the valence semiconductors turn out to be not so far away as all that from the free-electron sphere, quantitatively.

A second feature is the really remarkable empirical success in random alloys and in liquid metals — as Heine and Ziman (39) have emphasized — of what is called the "rigid band" model. This is the model in which one ignores the random background structure of the substance and pretends that the behavior is controlled by a sort of average band, which contains the average number of valence electrons that are contributed by each of the constituents. In the extreme form of the model one assumes that the properties of a metal are determined only by the number of valence electrons per atom, and not by which atoms contribute them. In some cases even crystal structures can be predicted by this kind of "averaging over the periodic table." A striking example is the alloy of Mo and Ru, which simulates the artificial element technetium in crystal structure and in a number of other properties, in all of which Mo and Ru differ markedly from Tc (40). More and more properties have recently been found to vary rather smoothly with this parameter, electron/atom ratio, which of course simply determines the size of the appropriate Fermi surface (10). I do not want to give the false impression that rigid bands always give a correct description of all properties of alloys and compounds — perhaps they work less often than not, especially when mixing widely different metals—but that it is rather a surprise that they work at all, in view of the sensitive balance of effects which determines most physical properties of metals.

2. Phillips, Cohen, and Heine have emphasized the relationship of the cancellation theorem to the observed occurrence of valence crystals as opposed to metals. In the lower-Z members of the periodic table there are many fewer core orbitals which one may use to expand the δ function in Eq. (18). This means that even the s part of the potential is less efficiently canceled in the lower elements, while in the first row there is no p cancellation at all, and d cancellation begins only toward the end of the second long period — from gallium on. Thus the remaining *effective* potential $(V + V_R)$ will be a decreasing function of the row number of the periodic table — from C to Si to Ge to Sn to Pb, for instance. Thus the earlier members will have stronger zone boundaries, wider gaps, and greater energetic advantages for forming an open valence structure. All these predictions are beautifully exemplified in these elements of the fourth column.

The variation with the second dimension of the periodic table, i.e., along a row, has not been as thoroughly discussed. I think this is a subject which is equally illuminated by the O.P.W. idea, and I should like to go into it briefly here.

The discussion will be based on the fact that there are actually two criteria determining whether the wave functions of the valence electrons in a solid may be reasonably well described as a set of orthogonalized plane waves with relatively small perturbation effects. The first we have already seen — it is that the core levels be quite deep, so that the orthogonalization leads to an effective potential which has its inner core reasonably well canceled out, and there is no strong admixture of core levels into the valence ones.

The second criterion may be understood by remarking that it is quite impossible energetically for the electronic configuration of an atom in a solid to be greatly different from that of the free atom. This is of course borne out very well experimentally by the X-ray structure factors of atoms in solids: They never differ by any serious displacements of more than one or two electrons from those of the free atoms. Some of you may have followed the discussion about X-ray measurements which seemed to indicate that iron was missing four d electrons in the solid; although the experiments did of course turn out to be wrong, this was at least the occasion for Herring's conclusive demonstration (41) that the energy required to move the four electrons was simply not available from the kinds of energies involved in cohesion. The displacement of two or three electrons by an appreciable amount compared to the Bohr radius a_0 — from one shell to another — costs rydbergs of energy.

Therefore a second important criterion is that *the atomic wave function we expect the valence electrons to have must be reasonably well describable by the occupation of plane-wave states within the Fermi sea of O. P. W. 's.*

It happens that atomic states are indeed not badly constructed to be approximated in that way under certain conditions. To see that this is so, we may look up [in Bethe-Salpeter, for example (42)] the Fourier transforms of atomic wave functions, i.e., their wave functions in momentum space. To give some examples, the 1s function is

$$\phi_{1s} = Ae^{-\kappa r}$$

and its (unnormalized) Fourier transform is

$$\frac{\kappa}{(\kappa^2 + k^2)^2}$$

that of a 2p function

$$\frac{\kappa^2 k Y_{1m}(k)}{(\kappa^2 + k^2)^3}$$

or of a 3d

$$\frac{\kappa^3 k^2}{(\kappa^2 + k^2)^4} Y_{2m}$$

A sketch of these functions is given in Figure 17. All of these fall off quite steeply with increasing k at approximately the points $k = \kappa$. If the Fermi surface is not too far from this point, a reasonable approximation to the atomic function may be achieved with O.P.W.'s.

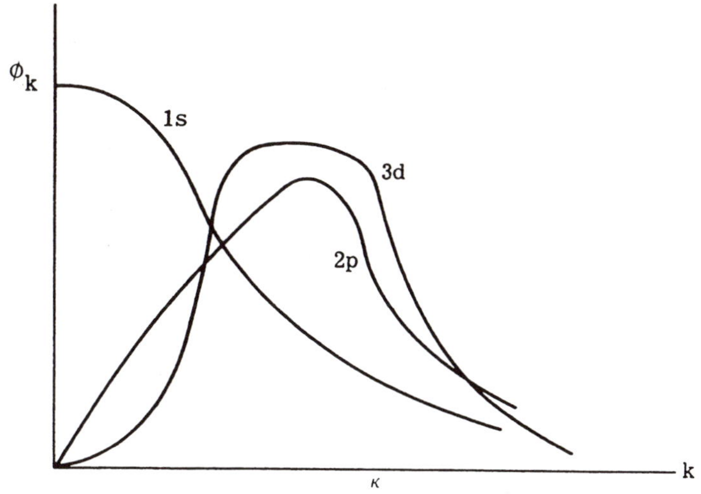

Figure 17

A second criterion for making such an approximation is easily visualized if we consider a Bloch wave made up out of atomic functions ϕ_{nl}. A reasonable approximation for such a wave might be

$$b_{k;nl} = e^{ik\cdot r} \sum_j \phi_{nl}(r - R_j)$$

One sees immediately that the momentum space representation of such a function is

$$b_{k;nl}(p) = \sum_k \delta_{p,\, k+K}\, \phi_{nl}(k + K)$$

It has the value $\phi_{nl}(k + K)$ at each of the points connected to k by a reciprocal lattice vector. Visualizing this in one dimension for, say, the ls function, we may sketch this function as shown in Figure 18. In momentum space we have a spike at each value of k + K, of height given by the size of the atomic momentum-space wave function at that point.

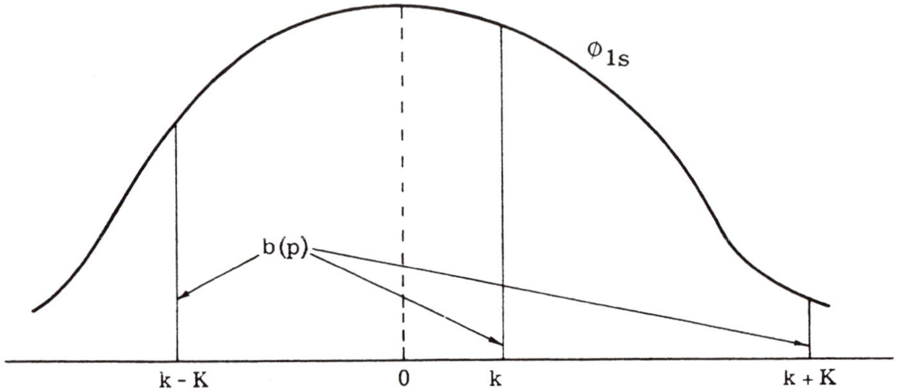

Figure 18

A wave function on the nearly free electron model consists of a single plane wave or, at best, a few plane waves of very nearly equal energy. The two kinds of wave function are then very dissimilar if, for instance, as in the above diagram, k and k - K both have large spikes but do not have very similar energy. That is, the K vectors must not be too small relative to k_F, which must be $\approx \kappa$. This is equivalent to saying that the important Brillouin zone boundaries must not be too far inside the Fermi surface, since the Brillouin zone boundary is precisely the point at which $|k - K|$ is equal to $|k|$, and, well beyond the BZ boundary, $|k| > |k - K|$ by a fair amount. There will be considerable tolerance in this requirement no doubt, but certainly when

$$|k - K| < \frac{3}{4} |k|$$

or so, there is no longer much resemblance between the two types of functions.

The physical sense of this requirement is just that the nearly free electron picture cannot describe a widely varying electron density very accurately, so that if the wave function size $(1/\kappa)$ is small compared to the interatomic distance (1/K) O.P.W. cannot be a good approximation. Another point of view is to say that when $1/\kappa$ is too small the remaining core potential begins to be too strong. The first requirement physically is equally understandable: The distribution of momentum in the atomic wave function and in the solid must cut off at roughly the same maximum kinetic energy; this requires $\kappa \approx k_F$.

The first of these requirements gives us a very straightforward criterion for "good metal" behavior; we have seen that the cutoff in the wave function and k_F must fall very near each other, and thus that the BZ must not be too far inside the Fermi sphere. But at least in Bravais' simple lattices we know that the first Brillouin zone contains precisely two electronic states per atom, so that if the number of valence electrons is Z,

$$\text{mean radius of BZ} = k_F \Big/ \left(\frac{Z}{2}\right)^{1/3}$$

(at Z = 2, vol. of BZ = vol. of Fermi sphere). Thus this is very directly a limita- on Z itself, and says that while a reasonable fit may be achieved for Z as large as 4 or 5, from then on the wave functions will begin to vary very rapidly in the radial direction and plane-wave models will not work.

It is instructive to see how these ideas work out along two typical rows of the periodic table: the second short period, Na, Mg, etc., and the first transition series, K, Ca, Si, Ti, ...

In the Na, etc., row, we are filling first the 3s and then the 3p levels. 3s and 3p atomic levels have radial nodes in their momentum space representations just as in their radial space functions, but the orthogonalization process will make precisely the corresponding node structure appear in the O.P.W. function; as a result we can think in terms of a "smoothed" space in which the radial nodes are ignored, and assume that the wave-function shape in this "smoothed" space is just that of a 1s or 2p function.

In 1930 Slater, generalizing from some Hartree calculations and various empirical data, produced a very successful set of approximate atomic wave functions and "screening constants" (43) for use in atomic and molecular calculations. These were smoothed nodeless functions of the form $r^{l'} e^{-\kappa r}$, and it seems that the appropriate κ to use in our calculation is probably just the κ from these formulas. These screening constants and wave functions have often been used as a pretty rough estimate of the potential and relevant wave function in the important outer region of the atom. We give here, for these elements, k_F calculated from the atomic volume in the metal, κ calculated as above from Slater, and also the quantity $\kappa_{\frac{1}{2}}$ — that k at which the atomic momentum functions have fallen to about one-half their peak value, Table 5. The agreement is not perfect (note that the errors in Mg and

Table 5

	Na	Mg	Al	Si	P
κ	1.39×10^8	1.80	2.20	2.60	3.00
$\kappa_{\frac{1}{2}}$	0.87	1.12 (p:1.6)	2.00 (s:1.4)	2.32	
k_F	0.90	1.36	1.55	1.84	

Al indicate a mixing of s and p character in these cases), but it is not bad at all; and it is amusing to see how nicely one may estimate the volume of a solid just by looking at the shape of atomic wave functions. P has various complicated crystal structures and does not resemble a metal at all, nor obey the $\kappa_{1/2} \cong k_F$ relationship. Thus P, S, and Cl cannot fit into the O.P.W. theory because the density at which the valence electron distribution would be reasonably uniform would be so high as to make the Fermi energy and thus the mean kinetic energy much too great relative to that in the atom. Another way to say it is that in the atom the existence of the shell degeneracy allows the filling up of a rather densely packed shell of electrons without excess kinetic energy, because the exclusion principle is satisfied by virtue of rather complicated phase relationships among the components of different momenta. This effect is important only in those elements with more than two or three electrons in a shell, but in these elements it prevents the formation of true metals, because in the true metal these phase relationships are necessarily destroyed and the kinetic energy required to maintain a given electron density near the core is much too high.

In P, S, Cl, as in N, O, F, the result is the formation of covalently bonded molecular systems, in which the p-shell holes very much retain their atomic p-function identity and form directed, saturated covalent bonds.

The same process but with a different result goes on in the transition series. Here, the electrons involved are orthogonalized to $(1s)^2$, $(2s)^2$, $(3s)^2$, $(2p)^6$, and $(3p)^6$. Thus they may take on 4s, 4p, and 3d character by virtue of the orthogonalization alone. O.P.W.'s with k's near zero cannot have appreciable d character and only weak p character, so that, naturally, at first, in K and even Ca, the O.P.W.'s are for practical purposes 4s wave functions. At higher k-values, however, a competition between 3d and 4p functions sets in. The 4s functions effectively screen the 4 p but not the 3d, and the higher n value means that $\kappa_{4p} < \kappa_{3d}$, so that at this point the density in k space of the 4p function is negligible, and the electrons — still essentially O.P.W.'s — begin to fill 3d levels. This may be shown by comparing screening constant κ's and the appropriate Fermi momenta, Table 6. Here, actually, the agreement is even better than before; and it is clear that Sc, Ti, and even V are quite satisfactory "O.P.W. metals." (The numbers in parentheses are calculated by supposing that toward the middle of this series the

Table 6

	K	Ca	Sc	Ti	V	Cr	Mn
κ	1.1×10^8	1.4	1.65	2.05	2.45	2.85	3.25 (κ_{3d})
$\kappa_{\frac{1}{2}}$	0.68	0.87	1.65	2.05	2.45	2.85	3.25 (κ_{3d})
					(2.2)	(2.65)	(3.22)
k_F	0.75	1.12	1.59	1.89	2.19	2.47	2.57

4s will begin to empty again.) At Cr and Mn, however, and to some extent in V, the lattice begins to be much less dense than required by the κ of the 3d electrons, and the "3d band" as an entity, separate from merely O.P.W.'s which happen to have 3d character near the nucleus, begins to take shape.

The behavior here is quite different from that in the p-shell series, however, because of the presence of the unfilled 4s-4p shell. This always retains some occupation, and as more and more d levels become occupied and the atoms relatively farther apart, the "O.P.W." Fermi surface loses the 3d electrons and returns to that appropriate for the 4s-4p electrons alone. Thus somewhere between V and Mn the d band proper, as a tight-binding band composed of electrons in approximately atomic states, takes shape; before this point, there is qualitatively no difference between a d band and any other band.

This appearance of the "inner shell" band at this point is marked by a great change in behavior, just as the failure of O.P.W. in the p shell was marked by the appearance of molecular and valence structures. Here the metallic character is retained, but the characteristic magnetic properties of inner shell electrons begin to make themselves evident — in actual ferro- and antiferromagnetism in the 3d row, in high magnetic susceptibilities and atypical behavior in the Pd and Pt groups.

In this digression I have tried to show why and how it is that holes in a nearly full shell and electrons in a nearly empty one behave entirely differently in their quantum chemical properties; and also where and why it is that the nearly free electron picture can no longer be expected to work. I do not believe such an analysis has been carried out before.

C. ONE-ELECTRON BAND THEORY IN THE PRESENCE OF PERTURBING FIELDS

1. Introduction

Before we leave the subject of single-electron theory it seems quite important to discuss the behavior of electrons in solids when some nonperiodic perturbation is added to the usual periodic field. Naturally, experiments on free electrons in perfect equilibrium in a solid cannot tell us much about them—at best, aside from the total energy, we may learn at most the density of states through specific heat or, hopefully, measurements of X-ray and optical spectra. In order to study the band structure we must perturb it. The perturbation may consist of externally applied electric and/or magnetic fields, which are for practical purposes constant relative to atomic dimensions, or of localized perturbing potentials caused by one or another disturbance of the periodic potential—impurities, dislocations, surfaces, etc.

Naturally, a truly weak and local perturbation, such as might be caused by hyperfine interaction with a nuclear magnetic moment, or by the difference between two isotopic nuclei, can be treated without appreciable complication by ordinary perturbation theory. A truly strong, local perturbation, such as might usually be caused by an interstitial or missing atom (vacancy) in the structure, is an intrinsically hard problem in a solid as elsewhere.

It happens, on the other hand, that the most common category of perturbations of interest in solids are those which, while they may be strong in the aggregate, are slowly varying relative to atomic dimensions. For instance, a uniform electric field leads to a perturbing term in the hamiltonian Eex, which even for small E can become as large as we like in a large crystal: $x \to \infty$. A uniform magnetic field is equally large numerically: the vector potential may be taken as $r \times H/2$, and grows linearly with the position vector r. Potential distributions caused by dislocations, charged impurities, excitons, surfaces, etc., also are often slowly varying but large. It is in this limit that the "effective hamiltonian" theory may be expected to be valid.

In this theory the assumption is made that the electron moves entirely among the states of a single band. We shall discuss the justification for this assumption later, but it is obvious enough that it is reasonable, in that a potential which is nearly constant and smoothly varying may move the electron adiabatically back and forth among states of nearly the same energy, but will not cause real transitions into other bands.

I shall first give a brief discussion of the dynamics of electrons within the "effective hamiltonian" or "effective mass" approximation, making the above assumption; then I shall go somewhat more deeply into some of the background of this approximation. A number of sources have fairly complete discussions of electron dynamics within it: Ziman (3) for electric and magnetic fields, Kohn in the Seitzschrift Vol. 5 for impurity states (44), Wannier (Chap. 6 of Ref. 4)—whose approach I shall more or less follow – and various others. Kittel has rather a nice qualitative discussion of the reason for "effective masses," which I recommend as collateral reading (2nd ed., p. 288) since I shall not be doing it that way. Wannier's book is reasonably good on the background justification, as good as any reference source except one which is difficult but complete: E. I. Blount in the Seitzschrift, Vol. 13.

It is remarkable that this problem has remained an active one for so long. The corresponding free-electron results were known by 1930—Landau's diamagnetism—and generalized by Jones and Zener, and by Peierls, shortly thereafter for, respectively, the electric and magnetic field cases, to "quasi-free" electrons in the effective hamiltonian approximation (46). Nonetheless, the first attempt at a respectable derivation of these results, and understanding of in what sense they were an approximation, came from Wannier's fundamental paper on the Wannier functions in 1937. The derivations in Seitz are quite incomplete. Again, the first breakdown of the approximation was discussed by Zener in 1934—the Zener effect—and Houston in 1940. It is, however, only with the work during the past decade of Luttinger, Adams, Wannier, Kohn, Blount, and Gibson that what one feels is probably a final understanding of the problem has been reached (45). It is somewhat surprising that all of this effort should be necessary to solve a single-particle Schrödinger equation of a particularly simple form, and the results are an object lesson in precisely how complicated in detail even the simplest phenomena of nature can be.

Let us then assume that the electron in the presence of a slowly varying potential does remain within a single band, the energy curve of which is $E_n(k)$—we discuss the nth band.

It is instructive to introduce immediately the concept of Wannier functions $a_n(r - R_j)$. These are functions obtained from a band of Bloch functions:

$$b_n(k, r) = e^{ik\cdot r} u_k^n(r)$$

by Fourier analysis in the following manner:

$$a_n(r - R_j) = N^{-1/2} \sum_k e^{-ik\cdot R_j} b_n(k, r) \tag{19}$$

Thus

$$b_n = N^{-1/2} \sum_{R_j} e^{ik\cdot R_j} a_n(R_j)$$

In a tight-binding band the a_n are obviously the original tightbinding functions, because then $u_k^n(r)$ is independent of k and $\Sigma_k \to \delta(r-R_j)$; otherwise the a_n are a properly orthogonal, normalized set of functions more or less localized around the sites R_j. In the single-band theory, the $a_n(r-R_j)$ represent the closest thing which can be achieved to eigenfunctions of position; they correspond to $\delta(r - R)$ for free electrons. Let us define an operator R such that $Ra_n(r - R_j) = R_j a_n(r - R_j)$.

Within the limitations that k occupies only the first zone and that R has as eigenvalues a discrete set of lattice positions, these represent a pair of variables conjugate to each other and resembling the p and r of free-electron theory, in that

$$[k, R] = 1$$

This is because — by definition —

$$i\frac{\partial}{\partial R} a_n(r - R_j) = i\frac{\partial}{\partial R_j} a_n(r - R_j)$$
$$= \sum_k kb_n(k, r)e^{ik\cdot R_j}$$

[by substitution of (19) for a_n]. Thus $i(\partial/\partial R)$ is equivalent to k, and vice versa.

Now if V(r) is slowly varying, we can approximate

$$V(r)a_n(r - R_j) \simeq V(R_j)a_n(r - R_j) = V(R)a_n(r - R_j)$$

In general, for slowly varying magnetic or electric perturbations, it is valid to assume that V(r) actually = V(R). Thus an approximate hamiltonian is

$$\mathcal{H} = E_n(k) + V(r) \simeq \mathcal{H}_{eff} = E_n(k) + V(R)$$

$$\mathcal{H}_{eff} = E_n\left(i\frac{\partial}{\partial R}\right) + V(R) \text{ or } = E_n(k) + V\left(-i\frac{\partial}{\partial k}\right)$$

Within the same limitations, it is obviously no more difficult to show that in the presence of a magnetic perturbation such that

$$E_{kin} \to -\hbar^2\left(\nabla - \frac{eA}{i\hbar c}\right)^2 \Big/ 2m$$

it is correct to replace p by $\hbar k$ and A(r) by A(R):

$$\mathcal{H}_{eff} \cong E_n\left(k - \frac{eA(R)}{\hbar c}\right) + V(R)$$

2. Weakly Bound Impurity States

The simplest of the three kinds of problems to which this formalism can be applied is that of a charged impurity atom in a semiconductor, such as P in Si (44). In Si P occupies substitutionally an Si site, but of course has an extra positive charge; it is a "donor." The result is that at sufficiently low temperatures an extra valence electron is bound in the region of the excess positive charge.

It happens that the resulting potential is rather weak, and the appropriate wave function rather long-ranged, both because Si has a rather large dielectric constant κ (the potential is $V = 2e^2/\kappa r$) and because of the low effective mass. Therefore one may use the "effective mass" or "effective hamiltonian" approximation. In the case of the valence electron in Si, the energy surface $E_n(k)$ consists of a set of six "valleys" at points $(k_0, 0, 0)$, where $k_0 \simeq 2\pi/a$ (0.85), along the six 100 directions. Near these valleys

$$E_n(k) \simeq \frac{\hbar^2(k_x - k_0)^2}{2m^*_{\parallel}} + \frac{\hbar^2(k_y^2 + k_z^2)}{2m^*_{\perp}}$$

Let us then write as a trial wave function in one of the valleys

$$f(r) = F(R)b_{k_0} = F(R) \sum_{R_j} e^{ik_0 \cdot R_j} a_n(r - R_j) \tag{20}$$

and let us note that

$$k_x b_{k_o} = i \frac{\partial}{\partial R_x} b_{k_o} = k_o b_{k_o} \qquad k_y b_{k_o} = k_z b_{k_o} = 0$$

Then

$$(k_x - k_o)^2 F(R) b_{k_o} = - \left[\frac{\partial^2}{\partial R_x^2} F(R) \right] b_{k_o}$$

$$- 2i \left[\frac{\partial}{\partial R_x} F(R) \right] k_o b_{k_o} + F(R) k_o^2 b_{k_o}$$

$$- 2k_o \left\{ - \left[i \frac{\partial}{\partial R_x} F(R) \right] b_{k_o} + F(R) k_o b_{k_o} \right\}$$

$$+ k_o^2 F(R) b_{k_o} = - \left[\frac{\partial^2}{\partial R_x^2} F(R) \right] b_{k_o}$$

so that b_{ko} may be factored out of the whole wave equation and we get simply an anisotropic hydrogen-atom wave equation:

$$- \frac{\hbar^2}{2m_{\parallel}^*} \frac{\partial^2 F(R)}{\partial R_x^2} - \frac{\hbar^2}{2m_{\perp}^*} \left(\frac{\partial^2}{\partial R_y^2} + \frac{\partial^2}{\partial R_z^2} \right) F(R) - \frac{2e^2}{\kappa R} F(R) = E F(R) \qquad (21)$$

The radius of this resulting wave function is something like $\hbar^2 \kappa / \langle m^* \rangle_{av} e^2 \cong$ about 50 Bohr radii in Si, κ being 12 and $\langle m^* \rangle_{av}$ [the appropriate average to use must come from a numerical solution of (21)] about $1/4 m_e$.

There are, in this approximation, six degenerate wave functions, one coming from each valley of the energy surface. At the central atom, the effective mass theory can never be really valid, and the failure of the theory causes a perturbation which splits up this degeneracy, with the lowest wave function being a symmetrized linear combination of all six, having therefore rather a high amplitude at the central atom and an energy rather far from the solution of (21). Kohn and Luttinger have considered the departures from effective mass theory in great detail. After correction for this effect, Kohn and Schechter have worked out the rather wild variation of the F(R) which comes from the interference among the six functions (20). Feher's classic experiments measuring the amplitudes at a variety of neighboring sites have given good qualitative but not perfectly quantitative support to this theory (47). Further detail on these effective mass wave functions is not suitable here.

3. Motion in External Fields

I shall follow Wannier in taking as the simpler example of the behavior and of the failure of the effective mass approximation in the presence of external fields the case of a uniform electric field rather than of a magnetic one. In an electric field, the effective hamiltonian is

$$\mathcal{H}_{eff} = E_n(k) - e\,\mathbf{E} \cdot R$$

Taking **E** in the x direction, this is

$$e_n(k) - eER_x$$

It is not very instructive to try to find explicit eigenfunctions of this hamiltonian, although it can in fact be done, because their interpretation would be rather meaningless, as we can see by realizing that in the free-electron case the electrons simply accelerate off to infinity.

Instead let us look at the equations of motion of the two canonically conjugate variables:

$$i\hbar \frac{dR}{dt} = [\mathcal{H}, R] = E_n(k) \left(-i \frac{\partial}{\partial k}\right) + i \frac{\partial}{\partial k} E_n(k)$$

or,

$$V = \frac{dR}{dt} = \frac{1}{\hbar} \nabla_k E_n(k)$$

This is a familiar result for the group velocity of a wave packet in a dispersive medium, but here the packet need consist, if we like, only of a single component of a definite momentum k, since $\nabla_k E_n(k)$ is a function of k alone. This relationship clearly doesn't depend on the form of the perturbation and is a general property of band electrons.

Next we find the equation of motion of k:

$$i\hbar \frac{dk}{dt} = [\mathcal{H}, k] = ieE \left(\frac{\partial}{\partial k_x} k - k \frac{\partial}{\partial k_x}\right) = ieE$$

Again, this is a remarkably simple result and can be immediately integrated: $k = k_o + Et/\hbar$. $\hbar k$ is often called the "crystal momentum P_c," and this equation states that $dP_c/dt = F$, where F is the force on the charge e. With these two equations, exact within the effective hamiltonian approximation, we can follow the motion of an electron starting in any individual Bloch wave $b_n(k_o)$, or of a wave packet centered about a given k_o.

Let us plot, then, the $E_n(k)$ curve along some line parallel to the x axis in k space; this will then give the variation of energy in time (Figure 19). Note that

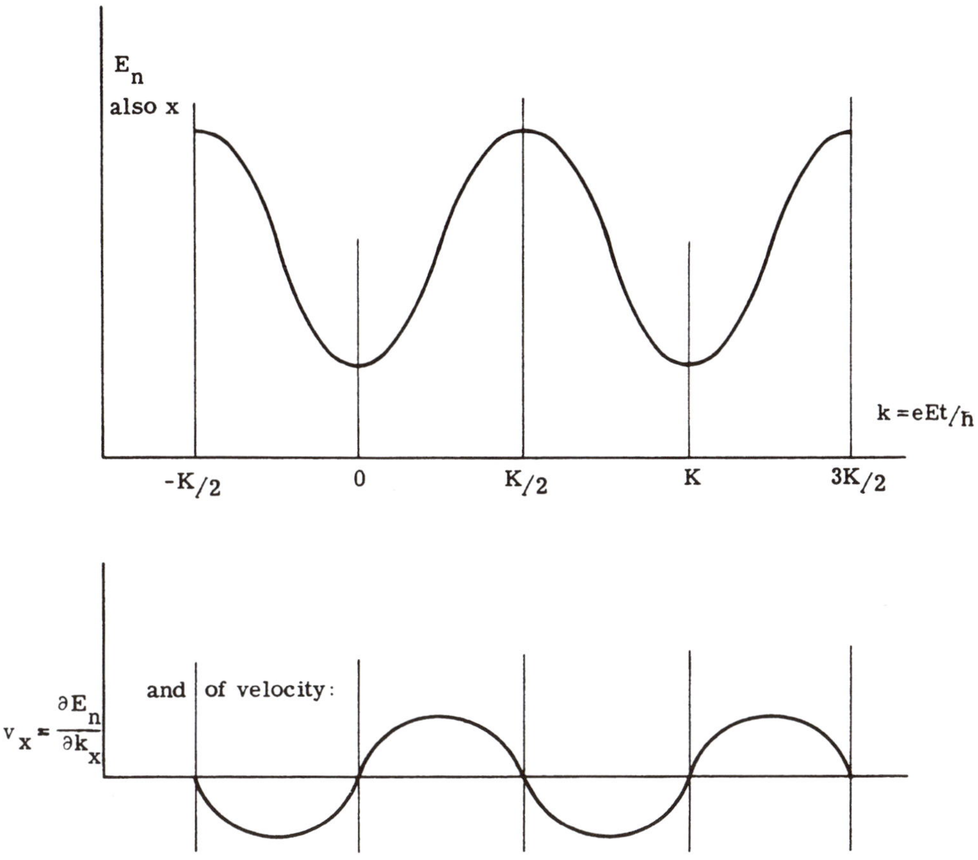

Figure 19

as k increases to the zone boundary we must make use of the periodicity of the energy bands in k space. We realize that as it passes the boundary it goes on into the next zone, which is just equivalent to hopping back to the corresponding point on the opposite face of the zone.

It is rather important to note that the resulting velocity is periodic, going through negative as well as positive values. The net displacement of a wave packet can be obtained by integrating the velocity against time:

$$x = \int_0^t dt \ v_x(t)$$

$$= \int_0^t \frac{dt}{\hbar} \frac{\partial E_n}{\partial k_x} = \frac{1}{\hbar} \int_0^t \frac{dt}{dk} \frac{\partial E_n}{\partial k} dk = \frac{1}{eE} \left\{ E_n \left[k_x(t) \right] - E_n(0) \right\}$$

Thus after each period the displacement returns to zero. This is not surprising when we realize that a full band of electrons must not carry any net current, since an insulator can be described by full bands.

The only reason why an actual metal or semiconductor carries a current is the presence of relaxation effects which stop the electrons from undergoing their periodic displacements as soon as they have accelerated only a tiny distance in k space: the resulting distribution is slightly increased in the +v direction, slightly decreased in -v, and the end result is a true current. On the other hand, in an insulator there are no empty final states available for scattering, so we must visualize the electrons as actually carrying out the periodic displacement.

The total displacement in the x direction is the band width W/eE. This encourages us to think in terms of a very useful semiclassical visualization of the band structure in the presence of an external potential (Figure 20). Semiclassically (not really classically, because the band concept is not classical) the electron of a fixed energy can be expected to oscillate back and forth between the upper and lower band edges, confined by the forbidden gaps.

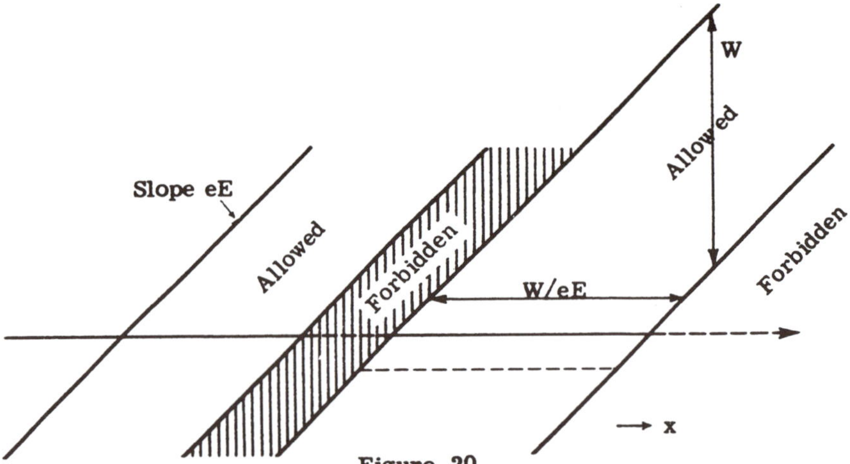

Figure 20

The total period in time of the oscillation is the time taken for d to move from one zone boundary to the next: i.e., $\hbar K/eE = T$. K is $2\pi/a$ (a = spacing of planes) so that the corresponding energy splitting is $h\omega = eEa$. The reason for this splitting is obvious: a wave function displaced by just a still satisfies the same wave equation, but the energy displacement is eEa. These "Stark ladders," as Wannier calls them (48), have actually been observed in p-n junctions. With normal electric fields attainable even in quite good insulators — 10^3 to 10^4 v/cm or 10 to 100 esu/cm — this splitting is extremely small — 10^{-6} to 10^{-7} ev — corresponding to very long periods, of the order 10^{-8} sec, as compared to mean free times for scattering of the order 10^{-11} at the very best. This is the reason why in the presence of scattering the distribution usually remains very close to equilibrium. In p-n junctions, on the other hand, and especially in tunnel diodes, one can obtain voltages of the order of the band gap over 10 to 100 lattice spacings, and small displacements may correspond to a reasonable energy (Figure 21).

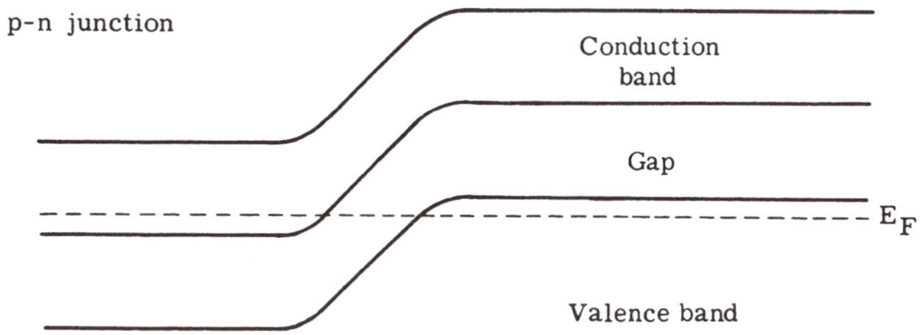

Figure 21

This energy splitting may also be calculated by a method much more analogous to the calculation of the splitting of the "Landau levels" which are so important in the magnetic case. Suppose we attempt to quantize the motion of the electron back and forth between the two band edges by the semiclassical phase-integral rule:

$$\int p\, dx = \hbar \oint k\, dx = nh$$

Let us now take the difference Δk between this and the next possible orbit:

$$2\pi = \oint \Delta k \, dx = \Delta E \oint \frac{\Delta k}{\Delta E} \, dx = \hbar^{-1} \Delta E \oint \frac{dt}{dx} \, dx$$

This leads simply to the Einstein frequency condition for the splitting:

$$\hbar^{-1} \Delta E = 2\pi/T$$

We have derived ΔE otherwise earlier, but in the magnetic case this method is the only simple one.

This then completes the discussion of the behavior of free electrons in homogeneous electric fields within the effective hamiltonian theory. As we see, the phenomena, while fascinating, are not very informative about band structure, but fortunately the case of the magnetic field is experimentally much more useful (50). I shall treat this very much more complicated case extremely briefly and then try to say a few words about where and how the whole effective hamiltonian theory fails.

The effect of a magnetic field is to replace the force eE in the crystal momentum equation $dp_c/dt = F$ by the Lorentz force $e \; (v/c) \times H$:

$$\frac{dk}{dt} = \frac{e}{c} \left[\nabla_k E_n (k) \times H \right] \tag{22}$$

Thus in the presence of a magnetic field **k** moves in a direction perpendicular both to H and to the gradient of the energy surface, which means it follows a contour of constant energy about the magnetic field (Figure 22).

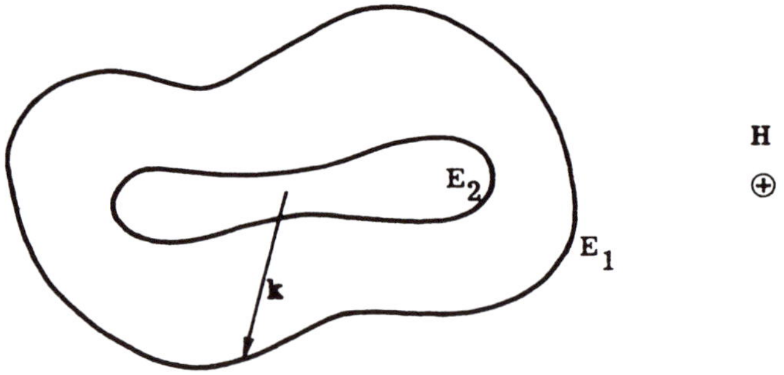

Figure 22

When $E_n = \hbar^2 k^2/2m^*$ Eq. (22) is easy to solve, giving the cyclotron frequency eH/m^*c. Otherwise it can be shown in various ways that the frequency is

$$\Delta E = \hbar\omega_c = 2\pi e/\hbar c \; \frac{\partial A}{\partial E}$$

where A is the area intersected by the appropriate plane normal to the field H. Correspondingly, there are quantized levels at those energies for which

$$A(E) = \frac{2\pi e}{\hbar c} n$$

These are the Landau levels, which are vital to the theory of the De Haas-van Alphen and cyclotron effects, which have been of great value in the study of actual band structures.

One important consequence of Eq. (22) has come into prominence in the past few years, particularly through the work of Lifshitz and his group in Kharkov. This is the possibility of so-called "open orbits" in the magnetic case (51). It is possible for a band structure to be such that as the electron drifts perpendicular to H along a constant-energy surface, this surface intersects the boundary of the zone, and the electron wanders off into the next zone and the next, and so on (Figure 23).

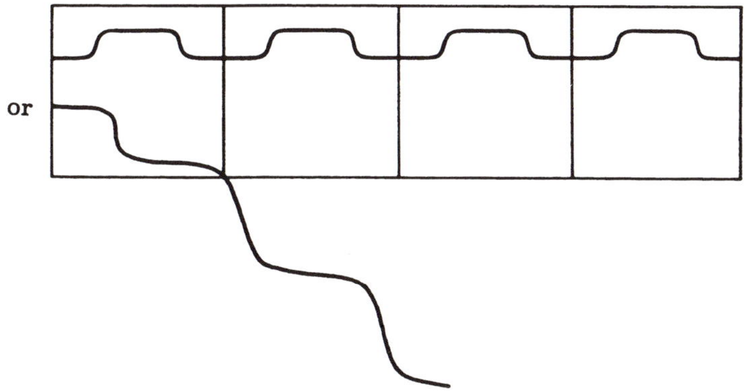

Figure 23

This orbit is very like that in the electric case; again, there is no net drift of the electron in the direction in which the orbit crosses the boundary, but it can drift in the other direction perpendicular to H, where electrons in the closed orbits cannot. Since $1/\omega_c$, as opposed to the electronic case, can be made short compared to τ, such orbits can have very strong effects. In particular, their existence has explained the decades-old puzzle of the nonsaturation of high-field magnetoresistance.

It should be emphasized that all of these results are qualified not only by those limitations of the effective hamiltonian theory which I shall shortly discuss, but also by the fact that in the magnetic case they are *only* correct semiclassically, unless the energy surfaces are, in fact, perfect spheres or ellipsoids. In the case of cyclotron resonance near a band-degeneracy point in Si and Ge, quantum deviations have been computed and measured for very low quantum-number orbits, by Fletcher and Luttinger (52, 53).

4. "Breakdown" Effects

One physically well-understood effect which lies completely outside of the domain of the effective hamiltonian is the Zener tunneling effect, which is not only of importance in practical applications of semiconductors but can be demonstrated quite easily on a very simple model. Let us go back to the perturbation theory for nearly free electrons and discuss what happens at a zone boundary if we now apply an electric field.

The equation $\hbar\,(dk/dt) = eE$ holds in any case, so that in the absence of the zone boundaries the electrons accelerate indefinitely in the field:

$$V = \frac{\hbar k}{m} = \frac{\hbar}{m}\left(k_o + \frac{eEt}{\hbar}\right)$$

In fact, we may find an exact solution of the time-dependent wave equation

$$i\hbar\frac{\partial\Psi}{\partial t} = -\frac{\hbar^2}{2m}\frac{\partial^2\Psi}{\partial x^2} - eEx\Psi$$

by setting $\Psi = a(t)\ \exp\left\{i\left[k + (Eet/\hbar)\right]x\right\}$

$$\left(i\hbar\frac{da}{dt} - eExa\right)\exp\left[ik(t)\right]x = \frac{\hbar^2}{2m}\left(k + \frac{eEt}{\hbar}\right)^2\Psi - eEx\Psi$$

where now a(t) satisfies a wave equation with a time-dependent kinetic energy

$$i\hbar\frac{da}{dt} = E^o\left(k(t)\right)a = \frac{\hbar^2}{2m}\left(k + \frac{eEt}{\hbar}\right)^2 a$$

This may be integrated but the result is quite irrelevant; the important thing is that by allowing k to vary with time the acceleration may be replaced by a time-dependent k vector and kinetic energy.

Now let us introduce the Fourier component V_K of the periodic potential. This will bring in a constant coupling in the wave equation, between k(t) and k(t) - K, which is another state also accelerating constantly. On an energy vs. k diagram, k - K moves along with k and we may plot it either on the same parabola or on one shifted by the reciprocal lattice vector K and crossing it at the zone boundary (Figure 24). Within the approximation that V is small, it is necessary only to consider the two k vectors which are near the crossing point P. Near this point we may expand the correct wave function as a sum of the two accelerated plane waves:

$$\Psi = a(t)\exp\left[i\left(k + \frac{eEt}{\hbar}\right)x\right] + b(t)\exp\left[i\left(k - K + \frac{eEt}{\hbar}\right)x\right]$$

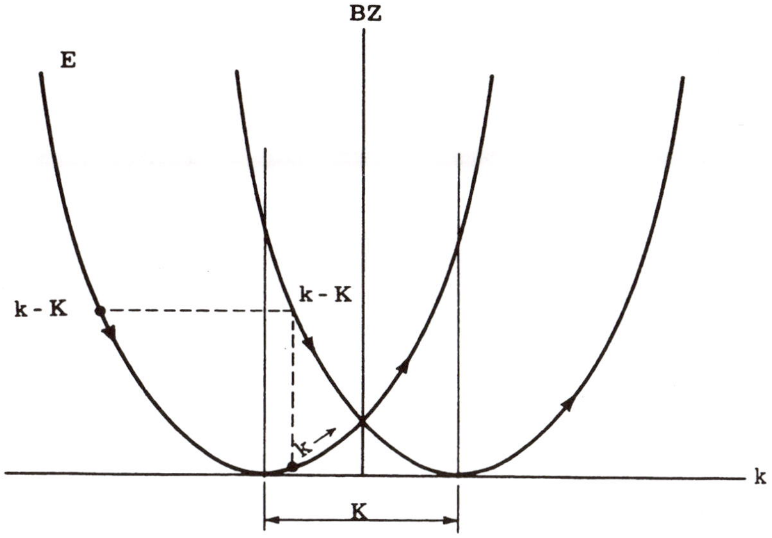

Figure 24

and it is easy to see that we get a pair of coupled equations for a and b:

$$i\hbar \frac{da}{dt} = E\,[k(t)]\;a + V_K\,b$$

$$i\hbar \frac{db}{dt} = E\,[k(t) - K]\,b + V_K\,a$$

The hamiltonian matrix is

$$\mathcal{H} = \begin{pmatrix} E\,[k(t)] & V_K \\ V_K & E\,[k(t) - K] \end{pmatrix}$$

This wave equation is of a form which is well known, in the theory of adiabatic or rapid passage, for instance. There are two limiting cases: If V is weak enough, it acts merely as a small perturbation. If we start with $a = 1$, $b = 0$, after crossing the zone boundary the coefficient a is still large, b is small, and the electron simply hops the energy gap and continues to accelerate. One can easily show that

$$b \simeq \frac{|V_K|^2}{\hbar \frac{d}{dt}(E_k - E_{k-K})}$$

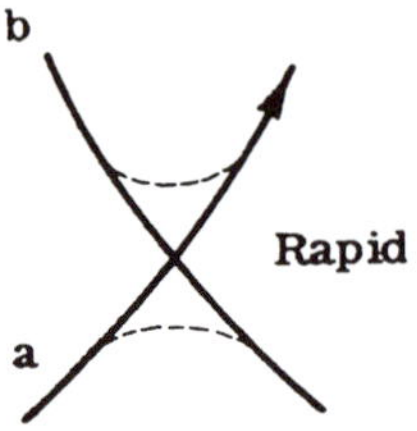

in this, which is called the rapid limit.

In the adiabatic limit, where V is quite large, it is more nearly correct to assume that the hamiltonian may be diagonalized at each instant of time, giving energies for the two states equivalent to the separate band energies $E_n(k)$. In this case a transforms *adiabatically* into b, while there is a very tiny probability of jumping the gap:

$$P \simeq \exp\left[-\frac{|V_K|^2}{\hbar \frac{d}{dt}(E_k - E_{k-K})}\right]$$

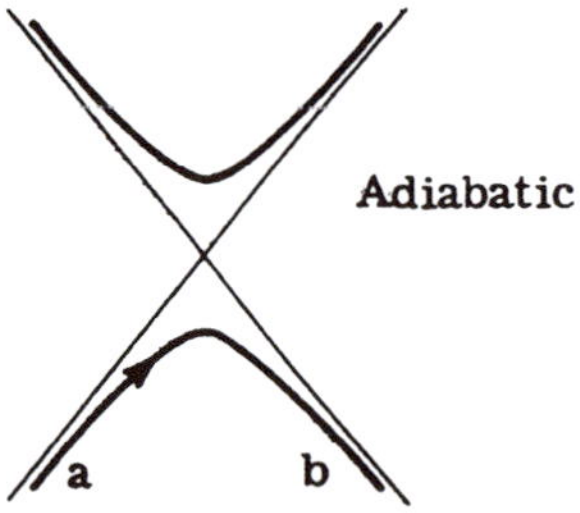

Now

$$\frac{d}{dt}(E_k - E_{k-K}) = \frac{2d}{dt}\frac{\hbar^2 k^2}{2m} = \frac{eE\hbar K}{m}$$

so

$$P \simeq \exp\left(\frac{-|V|^2 m}{\hbar^2 K eE}\right)$$

This is by far the more usual case. For a gap of 1 volt and $E = 10^4$ cm, $P = e^{-200}$. But when $E \sim 10^7$, as in a p-n junction, P is quite large and tunnelling is perfectly possible.

It was noted recently by Falicov (54), and a good theory was given by Blount (55), that a similar phenomenon occurs rather more easily in some crystals in the magnetic case. If a zone boundary interrupts a nearly free electron magnetic orbit, the electron as it runs around the energy surface can hop into the next orbit (Figure 25). As we have already noted the rate of motion in the magnetic case can become quite large, and the effect has indeed been observed in Mg. The effect is called "magnetic breakdown."

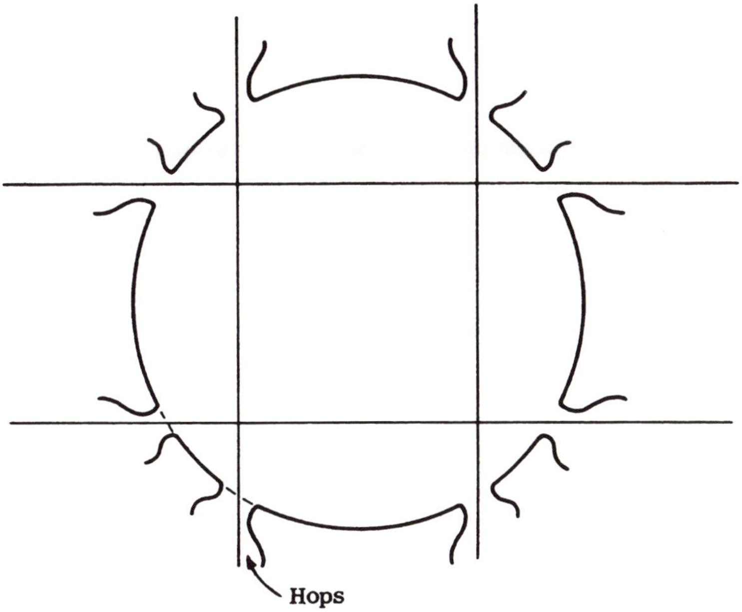

Figure 25

It is worth noting that in this limit E or H appears in the form exp - [const/E or H], so that these results could never appear as the power-series result of a perturbation theory in E or H, since all derivatives of $e^{-}(1/x)$ vanish at $x = 0$. We will find that the effective hamiltonian theory can be proved to be correct *to all orders* of a perturbation theory, but of course we have conclusively shown that it does not contain all physical effects nonetheless. Thus the perturbation theory can be asymptotically convergent at best.

5. Rigorous Basis of Effective Hamiltonian Theory

Now I should like to give a brief discussion of how one goes about improving on the effective hamiltonian theory (56). A very simple qualitative picture, due to Adams, will show the principle of this in a graphic way. Consider a very open lattice in which tight-binding theory is a rather good approximation, so that the Bloch functions are the tight-binding functions,

$$b_k^n = \sum_j e^{i\mathbf{k}\cdot R_j} \phi_n(r - R_j)$$

We are assuming that overlap is quite small, so that to a fair approximation the ϕ_n are also the Wannier functions $a_n(r - R_j)$. A picture of the real part of the Bloch wave function then might be as shown in Figure 26.

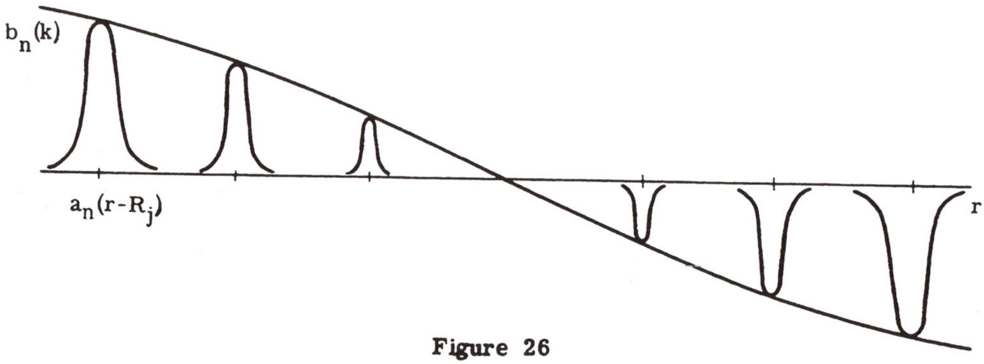

Figure 26

Now in deriving the effective hamiltonian we made the basic assumption that V(r), the true perturbation by the internal field, could be replaced by V(R), where V(R) is defined by the relationship

$$V(R)a_n(r - R_j) = V(R_j)a_n(r-R_j)$$

For a uniform electric field, for instance, this assumes that r may be replaced by its value at the center of the appropriate Wannier function. It is quite easy then to see that in the very localized case we are discussing, if $V(r) = -er$ is the linear potential as shown in this diagram, V(R) is the stair-step potential shown in Figure 27 and the difference is the sawtooth wave shown in Figure 28. Let us call this *difference* $X = r - R$. Obviously, in any physical case X will not look like the figure, but it can always be defined in terms of this difference, which in turn depends on $a_n(r - R_j)$. We notice immediately:

1. The difference X is small of the order eEa, where a is the lattice constant, so that the largeness of the perturbation is entirely contained in the operator R, and thus the effective hamiltonian theory contains essentially all the *large* effects.

2. X is periodic with the lattice period. Therefore to any order in perturbation theory we may redefine our Bloch and Wannier functions as the solution of a *new* band problem

$$\mathcal{H}b_k = \left(\frac{p^2}{2m} + V(r) - eEX\right) b_k \simeq (E_k{}^n)' b_k$$

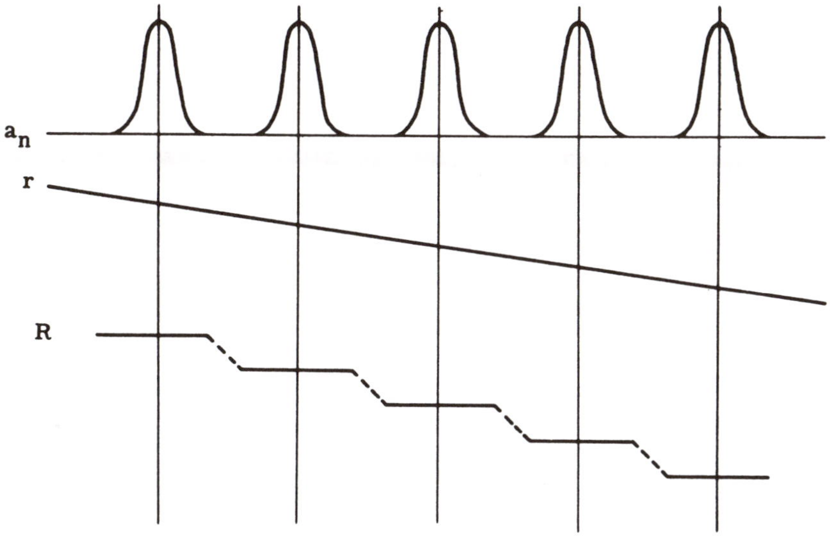

Figure 27

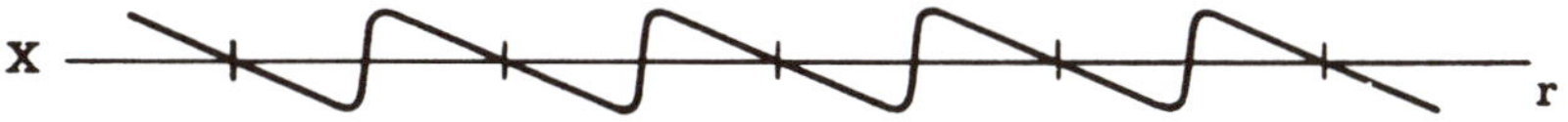

Figure 28

and thus eliminate the effect of X. Thus to any order in perturbation theory there is a new set of bands with new energy $(E_k^n)'$, new velocity (no longer given exactly by $\partial E/\partial k$), etc., and the effective hamiltonian theory is valid for the new bands with this proviso. Note however that X is *defined* in terms of a, which in turn depends on b_k, so that the only way to solve the equation is not once and for all but by iteration to all orders.

3. The effect of X on the Wannier functions in this case is in fact an obvious physical effect which was indeed left out of our original theory: an electric field not only must have the effect of accelerating electrons, it has the effect of polarizing the individual atoms' wave functions. For instance, it must mix in some of the p-band functions into the s band, etc., and this effect must be there no matter how localized the Wannier functions are. In other words, it is quite obvious that X may have interband matrix elements $X^{nn'}$; but we expect that these interband matrix elements will be those of a periodic potential, small of order eEa.

At this point this is merely a program, and it is the actual proof of these results which has occupied the group of people I mentioned over the past 10 years. To see what must be done, let us leave the tight-binding approximation and define X for a general band. There are actually two ways; the most obvious is in terms of Wannier function:

$$Xa_n(r - R_j) = (r - R_j)a_n(r - R_j)$$

so that the matrix elements of X are

$$(n, j\,|\mathbf{X}|\,n', j') = \int dr\; a_n(r - R_j)a_{n'}(r - R_j')(r - R_{j'})$$

Thus X is a function of $R_j - R_{j'}$, which proves it to be periodic but nonlocal.

The less obvious way is the so-called "crystal momentum representation." Clearly, since

$$b_k^n(r) = \sum_j e^{ik\cdot R_j} a_n(r - R_j)$$

$$Rb_k^n = -i\frac{\partial}{\partial k} b_k^n$$

But using the other representation,

$$-i\frac{\partial}{\partial k} b_k^n(r) = re^{ik\cdot r}u_k(r) - e^{ik\cdot r}\frac{\partial}{\partial k}u_k(r)$$

so that X is the operation $-i(\partial/\partial k)$ applied to u_k, the periodic part, alone. Again, this is clearly periodic, since the result is a Bloch function of the same k; but the operator is nonlocal. This crystal-momentum representation is more valuable in the magnetic case, because it is also necessary there to expand the operator p which appears in the perturbation in terms of periodic and nonperiodic parts.

In each of these representations there appears a possible difficulty. This is most obvious in the first: if the $a_n(r - R_j)$ are not sufficiently localized X may have large or even divergent matrix elements. For instance, the case that Wannier calculated in his original paper was that of free electrons considered as bands, and a_n fell off only as $\sin r/r$, so that the matrix elements would diverge rather badly.

The way around this is the contribution of Gibson and Blount. Notice that we defined

$$a_n(r - R_j) = \sum_k e^{ik \cdot R_j} b_k^n(r)$$

This is, however, by no means unique, because the different b_k's have an entirely arbitrary phase with respect to each other, since they are defined only up to an arbitrary constant factor. Thus we have an infinite space of possible a_n's:

$$a_n = \sum_k e^{ik \cdot R_j} e^{i\phi_k} b_k^n(r)$$

This phase transformation is, Blount showed, very like a gauge transformation, and most results are not really changed by it; but it is convenient to specify the phases by the requirement that the a_n be *as localized as possible* — by a variational principle, that is.

We can see precisely the same trouble in the $\partial/\partial k$ definition; again, each b_n may be multiplied by an arbitrary phase, and it is only by some particular specification that we can make X a smooth function of k.

In the case of a single, separate band a_n may be proved to fall off exponentially with distance in the best phase specification, and the whole theory is very well convergent. When lines or points of degeneracy occur, the a_n do not fall off very rapidly, although they do, in fact, fall off just rapidly enough for convergence. But much simpler results may be obtained at such a point or line by making new linear combinations among the different bands in such a way that the Wannier functions are well localized and the X and k operators well-behaved. Then, however, the unperturbed band energy is no longer a numerical function $E_n(k)$ but a matrix $E_{nn'}(k)$. This procedure was first used by Luttinger in his treatment of cyclotron resonance at the degenerate valence-band maximum in Ge. Using this technique of phase specification, Blount has proved the validity of the effective hamiltonian theory to all orders of perturbation theory.

It might be worthwhile to comment without further detail on the rather fascinating complexity of some of the results one can obtain. One important result is the fact that to various orders in E or H new and rather surprising terms in various physical operators enter — for instance, the famous anomalous current in the anomalous Hall effect (57) — which can lead to currents, energy transports, etc., which are purely quantum polarization effects outside the domain of normal transport theory. Many other effects, complicated but interesting to the specialist, appear when such further complications as spin-orbit coupling are properly introduced.

3

ELEMENTARY EXCITATIONS

A. THE IDEA OF ELEMENTARY EXCITATIONS: GENERALITIES ON MANY-BODY THEORY

Everything I said during the first term of this course was based on an approximation — an approximation, in fact, which we can easily convince ourselves is a very rough one in most cases. That was the approximation that the electrons in the solid move independently of each other, and of the motions, which we did not even discuss, of the heavy ion cores. At the very best, we have taken into account the interaction of the electrons — except for a brief reference to the Wigner-Pines calculation of correlation energy — in terms only of its average, i. e., of Hartree-Fock theory.

The electrons and ions in a solid do, in fact, interact very strongly, and in what would be expected to be a rapidly fluctuating way. Two simple instances are worth noting. We saw that the correlation-energy correction was comparable to the binding energy of the alkali metals, and thus could hardly be considered to be a small effect. We also note that the interaction energy of two charges at fairly large distances must be reduced by the dielectric constant κ: $e^2/r \rightarrow e^2/\kappa r$. Values of κ for typical solids range in the order of 3 to 20 even for insulators, so that the resulting change in the interaction is rather large.

Both are true many-body effects, lying entirely outside the domain of Hartree-Fock theory. What I shall be doing for the remainder of these lecture notes is to discuss these many-body effects. Unfortunately, no good text on any but the highest level seems to exist in this field, even though many of the concepts are not difficult; perhaps the best is Pines in a set of lecture notes and reprints (58). Other references are Landau-Lifshitz (Chapter VI of Ref. 6) and, more advanced, but the most readable of the highbrow books, Thouless, "Quantum Mechanics of Many-Body Systems" (59); also the Les Houches notes (80).

How can it be then that the use of simple one-electron concepts leads us so often to qualitatively, and even sometimes semiquantitatively, correct conclusions? There are actually quite a number of answers to this question, with only one of which we shall be primarily concerned in these next few lectures. Nevertheless it might be well to list several of them here: (1) the variational theorem, (2) the exclusion principle, (3) screening, and (4) the concept of elementary excitations.

1. The Variational Theorem

The first, the variational theorem, almost goes without saying — but it applies to the ground-state energy only. That is, although the Hartree-Fock wave function is not a very accurate approximation to the true wave function of a system containing many electrons and ions (for a large system, there is almost no overlap between the two), nonetheless its energy is not anywhere nearly so bad an approximation.

2. The Exclusion Principle

In the Hartree-Fock theory of electronic systems the exclusion principle plays a very important role in reducing the many-body effects. While recognized many years ago for the free electron gas, this effect has been pin-pointed and formalized, particularly by Weisskopf (61), and later by Brueckner in the case of nuclear particles; and a similar effect for finite electronic systems has been particularly emphasized by MacWeeny (62).

In Section 2-A1 we wrote the Hartree-Fock Wave function for a many-electron system in the following way:

$$\Psi_{HF} = \prod_{\substack{\text{occ. states} \\ n,\sigma}}^{N} c_{n\sigma}^{\dagger} \Psi_{vac}$$

where the $c_{n\sigma}^{\dagger}$ create electrons in a set of orthogonal orbitals $\phi_{n\sigma}$, which are the best single-electron orbitals of the problem, in the sense that we eliminated all matrix elements having the effect of exciting one single electron alone out of an occupied state $\phi_{n\sigma}$ into an unoccupied one $\phi_{M\sigma}$, $M > N$. This, as we explained, was the sense of the equation $\left[\mathcal{H}, c_{n\sigma}^{\dagger}\right] = E_n c_{n\sigma}^{\dagger}$ + truly off-diagonal terms.

Then any perturbation which can occur must simultaneously take two electrons out of occupied states n, n' into unoccupied states m, m'; the appropriate matrix elements are

$$\left(mm' \mid e^2/r_{12} \mid nn'\right)$$

$$= \int dr_1 \int dr_2 \frac{2}{|r_1 - r_2|} \phi_m^*(r_1)\, \phi_{m'}^*(r_2)\, \phi_n(r_1)\, \phi_{n'}(r_2)$$

For instance, the lowest-order correction in perturbation theory to the energy is

$$-\sum_{\substack{n,n'<N \\ m,m>N}} \left|\left(mm'\,|\,e^2/r_{12}\,|\,nn'\right)\right|^2 \Big/ \left(E_m + E_{m'} - E_n - E_{n'}\right)$$

Because of the exclusion principle, these matrix elements are likely to be rather small. The reason is that the states m and m' must be chosen from among a subspace of states which are *orthogonal* to the states n and n', and are likely to be very different from them — either to occupy different places in real space, or to vary rapidly relative to them, i. e., to be in different parts of momentum space. Thus the integral is usually unexpectedly small. A good way to put it is to point out that the integral is the Coulomb interaction energy of the charge distribution

$$\rho_{mn} = \varphi_m^\dagger(r_1)\,\varphi_n(r_1) \text{ with } \rho_{m'n'} = \varphi_{m'}^\dagger(r)\,\varphi_{n'}(r)$$

But, by orthogonality,

$$\int dr\,\rho_{mn}(r) = \int dr\,\rho_{m'n'}(r) = 0$$

i. e., these charge distributions alternate in sign, usually rather rapidly, so that their interaction is not very great. This is MacWeeny's argument.

The nuclear physicists put it in the form that electrons n and n' are very limited in the amount of phase-space they can scatter each other into by the exclusion principle. For instance, in the free electron gas they have to scatter each other entirely out of the Fermi sphere, and this weakens the scattering (Figure 29). We shall return to this topic both in the study of Fermi liquids and of the magnetic state.

Note that the scattering of each of the possible pairs n, n' will be finite, so that the true Ψ_g will have a finite modification from Ψ_{HF} for each n. Thus as $N \to \infty$,

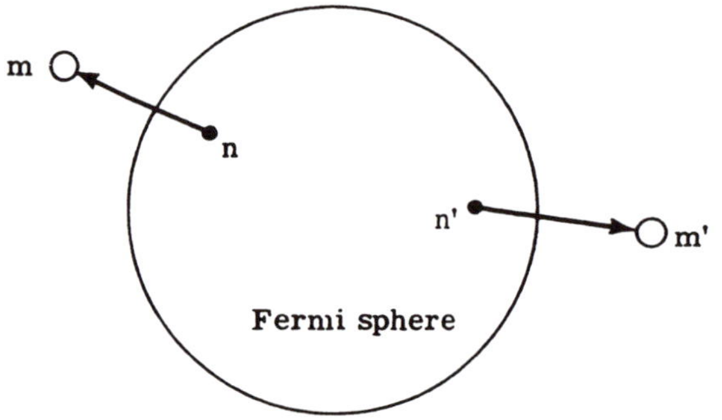

Figure 29

since Ψ_{HF} is a product of N factors, each of which has been modified by a finite amount, its overlap with the true $\Psi_g \rightarrow 0$.

3. Screening

When we introduce a charge — as, for instance, the charge of a particular electron whose motion we may be studying — into a gas of electrons, the other electrons will, of course, be attracted by it and move toward it if it be positive, be repelled and move away if it is negative. In the latter case, for instance, the motion leaves behind a deficiency of electronic charge relative to the original Hartree-Fock density, which is equivalent to a certain amount of positive charge. The external electron thus interacts with other electrons at any distance as though it had a smaller charge — it has been *screened* by the other electrons. This acts very strongly to reduce correlation effects.

4. The Concept of Elementary Excitations

The final idea, and one which I shall be speaking on for some pages, is the idea of of elementary excitations. This concept again is an idea that has grown rather gradually, so gradually indeed that one can hardly name an originator, although perhaps Landau [in 1941 (63) and his book (6)] more than anyone else is responsible for the conscious realization that a general approach of this sort existed. He introduced the name "elementary excitations." Fröhlich (64) and Pines (65) are responsible for the realization of the value of exploiting the methods of quantum field theory, which has opened up the whole subject of many-body theory as a mathematical discipline; and Kohn (66) and Luttinger (67), along with a number of others but more than anyone else, for the concept that rigorous proofs of some of the ideas of elementary excitation theory are possible. But one must not leave out the name of Debye, as the first user of an elementary excitation theory (Debye phonon theory), or of Wigner, as the first to reason on the subject of correlated electronic motions.

There are two ideas behind the concept of elementary excitations. First is the idea that the total binding energy of the ground state is not a very important physical quantity, and does not have much to do with the behavior of a physical system. What *is* important physically is the behavior of the lower excited states relative to the ground state: those states, that is, which are likely to be excited at relatively low temperatures or by weak external fields. We think immediately of a metal or a semiconductor, in which all the behavior is determined by the low excited states which we speak of as having a few moving charge carriers, or of the elastic or thermal properties of a solid, determined by the presence of a small number of lattice waves, which we call phonons. Thus our interest is often focused on the set of low-lying excited states of a system as the physically most fundamental property of it.

The second idea is that the low-lying states often — in fact almost always — are of a particularly simple character, and can be treated with much greater mathematical rigor and physical understanding than other states. Let me explain the reason for this.

As I have been emphasizing, in spite of the minimizing effects we have been talking about, the many-body corrections to the wave function and energy of the true ground state Ψ_g are very great: in particular Ψ_g and the single-particle approximation Ψ_{HF} have little or nothing to do with each other, because Ψ_g differs in the coordinates of each of the N particles.

Now we do know, usually, something about the group properties of the ground state. In an insulator, for instance, the electronic wave function has the full symmetry of the translation group: $T\Psi_g = \Psi_g$.

For simplicity, let us consider excited states of the system with precisely one added particle. In Hartree-Fock approximation, the lowest such state of a given crystal momentum k would be the state in which the extra electron is in the lowest empty band in the state of momentum k:

$$\Psi_k = (c_k^{n'})^\dagger \ \Psi_{HF}$$

This is of course nothing like an eigenstate. We can, however, compare the lowest eigenstate Ψ_k of momentum k and one extra electron to the *true* ground eigenstate Ψ_g. There will be some operator $q_k^\dagger$ which represents the relationship between these two states:

$$\Psi_k = q_k^\dagger \ \Psi_g$$

The most important thing about the operator $q_k^\dagger$ is that it inevitably represents only a very small displacement of the entire system. In the simple case we are discussing, we shall see that in fact out of the $N \sim 10^{23}$ electrons only those of momentum k are disturbed a finite amount. In more complicated cases, it may be that every electron is displaced in some sense—as in a plasma oscillation—but only by an infinitesimal amount.

Thus, as far as almost all the electrons are concerned, Ψ_k is the same, nearly, as Ψ_g. If, for example, we consider a wave packet made up from k's near a central value k_o,

$$\Psi_{packet}(k_o, r_o) = \int \exp\left[-ik\cdot r_o - \frac{1}{2}\left(\frac{k - k_0}{\Delta k}\right)^2\right] \Psi_k \, dk$$

we can expect that the resulting disturbance in the wave function will be localized to the region near r_0 of size $1/\Delta k$. If the wave function were truly Hartree-Fock, of course, this would necessarily be so: Ψ_{packet} would be

$$(\Psi_{\text{packet}})_{HF} = \int e^{-ik\cdot r_o} \exp\left[-\frac{1}{2}\left(\frac{k-k_o}{\Delta k}\right)^2\right] c_k^\dagger \, \Psi_{H-F}\, dk$$

$$= \int dr \left(\int e^{-ik\cdot(r_o - r)} \exp\left[-\frac{1}{2}\left(\frac{k-k_o}{\Delta k}\right)^2\right] dk\right) \psi^\dagger(r)\, \Psi_{HF}$$

$$= \left(\int dr\, f_{\text{packet}}(r)\, \psi^\dagger(r)\right) \Psi_{HF}$$

where

$$f_{\text{packet}}(r) \propto e^{ik_o\cdot(r-r_o)} \exp\left[-\left(\frac{\Delta k}{2}\right)^2 (r-r_o)^2\right]$$

In the more general case, we feel that the formation of a packet will result in a disturbance of the true ground state Ψ_g which is localized. Thus the operator

$$q^\dagger_{\text{packet}}(r_o, k_o) = \int dk \exp\left[-ik\cdot r_o - \frac{1}{2}\left(\frac{k-k_o}{\Delta k}\right)^2\right] q_k^\dagger$$

is an operator which creates only a localized disturbance.

If, then, we form a wave function which has two of these wave packets present

$$\Psi(k_o, r_o; k_o', r_o')$$

$$= q^\dagger(k_o, r_o)\, q^\dagger(k_o', r_o')\, \Psi_g \qquad k_o \neq k_o', \quad r_o \neq r_o'$$

we can expect that the two packets will not interfere with each other very much, for most of the possible values of r_o and r_o'. Thus if the excited state Ψ_k has excitation energy $E_k = (\Psi_k|\mathcal{H}|\Psi_k) - E_g$, we can expect that the excited state containing two such excitations has the sum of the two energies, to order $1/N$ (that being the order of the actual overlap of the two wave functions).

$$E(k_o, k_o') = E_{k_o} + E_{k_o'} + O\left(\frac{1}{N}\right)$$

This is the fundamental thought behind the concept of elementary excitations: that in a very large system two such excitations will not interfere, whether they be simple quasi-particles such as I have discussed here, phonons, excitons, or something even more complicated. Thus, when the number of excitations present is sufficiently small, the properties of the system, and particularly the energy, will be a linear superposition of the properties of noninteracting elementary excitations. Another consequence in this case is that two q_k's which are not interacting will have the true fermion anticommutation rules. It is not easy to show, but it is true, that we can always take the q's as having fermion commutation rules in general.

This, then, is a preliminary definition we can make of an elementary excitation: *An elementary excitation of momentum k is that operator which creates the lowest excited state of a particular type of momentum k from the ground state.* From the above reasoning, we expect these excitations to interact only weakly— in order 1/N, in fact—so that the system can contain a relatively large number of them, and thus be in a state of a very high degree of excitation in terms of absolute number n of excitations, and still we can treat the excitations as an approximately noninteracting gas of independent particles. It is important to realize that the above idea of packet formation is not confined to the particular type of excited state I discussed, but may be applied to lowest eigenstates chosen in such a way as to represent single phonons, spin waves, etc. Another way of saying it is that the concept of elementary excitations is a way of *linearizing* the equations of the system about the true ground state rather than about some independent particle approximation.

Before going further into qualifications and extensions of the above definition, it might be very well to give some concrete examples. In addition, I should like to support the statement I made earlier that our physical classifications of solids and of some liquids are based to a large extent on the types of elementary excitations which they exhibit. Thus I shall write the elementary excitations and the types of material in which they are important in list form, with the elementary excitation on the left and the type of material it characterizes on the right.

1. Free electronic charge carriers

a. With energy gap, energy minimum near isolated points in momentum space	Insulators and semiconductors
b. With energy gap, energy minimum at a surface in momentum space; charge state peculiar	Superconductor
c. No energy gap; E_k vanishes at a surface in k space	Metal (normal)

Note: Polarons are merely a further elaboration of all the above, except for the self-trapped polaron, which may be the dominant excitation in certain substances which are neither metal nor insulator.

2. Density waves

a. Longitudinal phonons	Solid; quantum Base liquid; certain cases of quantum Fermi superfluid
b. Transverse phonons	Solid
c. Zero sound	Some quantum Fermi normal fluids
d. Spin.density waves	Some quantum Fermi normal fluids
e. Plasmons	Metals and many insulators — most charged quantum fluids

Note: Ordinary sound in classical fluids and second sound are *not* elementary excitations in the above sense, but purely statistical phenomena. They depend for their existence on the establishment of local equilibrium by irreversible processes.

3. Spin fluctuations

a. Ferromagnetic spin waves: no gap, $\Delta m = 1$	Ferromagnets
b. Antiferromagnetic spin waves: gap, $\Delta m \to 0$ as $k \to 0$	Antiferromagnets
c. Spin rotations in general; k not defined, but still an important example of an elementary excitation theory	"Magnetic state"—ferro-, antiferro-, and paramagnets

4. Miscellaneous

a. Excitons, including spin excitation waves of Slater	A two-particle excitation of insulators, and perhaps superconductors
b. Rotons (energy gap, surface in k space)	Liquid He_4
c. "d-mons"	(A density wave of questionable existence in substances containing two or more types of carrier with widely different masses)
d. Various excitations associated with imperfections in the periodic structure — surface states, localized modes, impurity states, bound excitons, etc. These are a matter for the student of details — possibly crucial in the behavior of real solids, but not of much basic importance in the m.b.p.	

As you can see, a majority of the interesting phenomena of solid-state physics find themselves categorized somewhere in this list. Not all however; not even, actually, all the entities which have in common with them the character of small, elemental disturbances of the regular structure of the normal solid, which in many ways we treat as independent particles. Such entities as vacancies, interstitials, and impurities seem to deserve a separate category of "elementaryness" which is not very similar to the elementary excitation idea, primarily in that their behavior is in a real sense classical rather than quantum mechanical. Dislocations, and their "quantum" counterparts, quantized vortices and flux lines, are still another category; these are much larger (size $\sim N^{1/3}$) and more complicated phenomena.

In the case of each of the elementary excitation types, the approximate independence mentioned above allows us a common approach to the calculation and the understanding of thermal and transport properties. We treat the elementary excitations as, to a zero'th approximation, a gas of noninteracting Bose or Fermi particles, as the case may be, and then the statistical mechanics and transport theory become no more difficult than they are for a perfect noninteracting gas. In more sophisticated applications, we can include the interaction of the elementary excitations to any desired order; the lowest order will allow them to scatter on each other when they approach within a short distance. The interaction which scatters the elementary excitations, like the excitations themselves, must be considered as the result of an actual calculation which necessarily brings in the N-body character of the system, but in many cases we can treat it phenomenologically or approximately and achieve reasonable results. Finally, in a number of important cases we can actually draw conclusions about the nature of the ground state itself by reasoning backward from the properties of the elementary excitations, as in the Debye-Waller theory of zero-point motion of phonons.

B. THE N + 1 BODY PROBLEM

1. Quasi-Particles

Let us talk about the particularly simple example of elementary excitations with which we started this discussion, namely, free carriers in a semiconductor or insulator. This problem is often called the "N + 1" body problem because we are really interested only in the changes which come from the addition of the extra particle. There are two general types of elementary excitations; this is an example of the type which are called "quasi-particles," because they closely resemble particles in independent Hartree-Fock single-electron wave functions and because they have the same commutation rules and charge as the individual particles in the medium from which they spring — in the usual case, they are fermions. The contrasting type of elementary excitation is the collective excitation, which includes phonons, plasmons, excitons, spin waves, etc. Rotons in He_4 are an anomalous case and share the properties of each type.

The characteristic of a quasi-particle excitation $q_k^\dagger$ is that it has a finite overlap with the corresponding single-particle approximation $c_k^\dagger$:

$$(q_k^\dagger \Psi_g, \; c_k^\dagger \Psi_g) \;=\; Z^{1/2} \tag{1}$$

Z finite (not of order 1/N). This must be thought of as a definition; we cannot prove that such excitations always exist, nor should we be able to, because there are systems in which for practical purposes they do not. Equation (1) is not trivial, because we know of course that the overlap between the actual ground state Ψ_g and the Hartree-Fock approximation Ψ_{HF} is of order e^{-N}. Thus in discussing quasi-particles we are working with a concept removed only a finite distance from the independent-particle model, which is a great conceptual advantage.

One way of thinking of Eq. (1), which corresponds very closely to the way the many-body theorists actually deal with quasi-particles (68), is to consider the question: suppose at time t = 0 I were to operate on the true ground state with the operator $c_k^\dagger$ which creates a particle in (say) a plane-wave state $e^{ik\cdot r}$. Now I wait a very long time T, during which time the hamiltonian operates on the system, so that the new wave function at time T is $e^{-iHT} c_k^\dagger \Psi_g$. Now, I ask if the particle is still in the state k. That means I apply the destruction operator c_k and take the scalar product with the original ground state. Thus I compute

$$\begin{aligned} G_k(T) &= \left(e^{-iHT}\Psi_g, \; c_k e^{-iHT} c_k^\dagger \Psi_g\right) \\ &= \left(\Psi_g, \; e^{+iHT} c_k e^{-iHT} c_k^\dagger \Psi_g\right) \\ &= \langle g |\, c_k(T)\, c_k^\dagger(0)\, | g \rangle \end{aligned} \tag{2}$$

Now, to evaluate this, it is useful to insert into this matrix element a set of exact excited states, one of which is the quasi-particle state $q_k^\dagger \Psi_g$ but of which, as we have said, there may be a very large number. For instance, it is very likely that $c_k^\dagger$ can have the effect of making a state with two quasi-particles and one quasi-hole, the sum of the momenta of the three adding up to k: $q_{k_1}^\dagger q_{k_2}^\dagger q_{k_1+k_2-k} \Psi_g$. Let us call any one of the very large number of such possible excited states Ψ_n, and note that almost all the possible types of Ψ_n will enter not with finite but with infinitesimal amplitude, because there an infinity of such states with the same momentum. (The only possible exception, which is trivial to deal with, is other band states in a periodic potential.) Thus

$$G_k(T) = \sum_m \langle g|c_k|m\rangle \, e^{-i(E_m - E_g)T} \langle m|c_k^\dagger|g\rangle$$

$$= Z e^{-iE_k'T} + \sum_n |\langle g|c_k|n\rangle|^2 \, e^{-iE_n'T}$$

Here the excitation energy E' has been defined as $E - E_g$. The sum over n will be an integral in the limit $N \to \infty$, of the form

$$\int e^{iE_nT} \, dE_n \; f(E_n)$$

where all the E_n's are greater than E_k. Such integrals always lead to rapidly decaying functions; thus G(T) at large times is controlled by the one quasi-particle excitation. We can say that, having created the electron in state k, an amount Z of it remains after a very long time, the rest having decayed away into a continuum of states; this amount Z behaves as if it had the energy of the quasi-particle.

Note that this gives us an "operational" way of defining the quasi-particle operator q_k. That is, we may use the equation

$$\left(c_k^\dagger \Psi_g\right)(T) = e^{-iHT} c_k^\dagger \Psi_g$$

$$= Z^{1/2} \, e^{-i(E_k' + E_g)T} q_k^\dagger \Psi_g + \int dE_n \, e^{-i(E_n' + E_g)T} f(E_n) \Psi_n$$

multiply this by $e^{i(E_k' + E_g)T}$ and integrate to get

$$q_k^\dagger \Psi_g \propto \int_0^\infty dT \; e^{i(E_k' + E_g)T} \left(c_k^\dagger \Psi_g\right)(T)$$

or

$$q_k^\dagger = \text{const.} \int_0^\infty dT \; e^{i(E_k' + E_g)T} \; c_k^\dagger$$

That is, we can select out of the time-dependent wave function the part belonging to the true quasi-particle just by appropriately averaging over all times, which gives

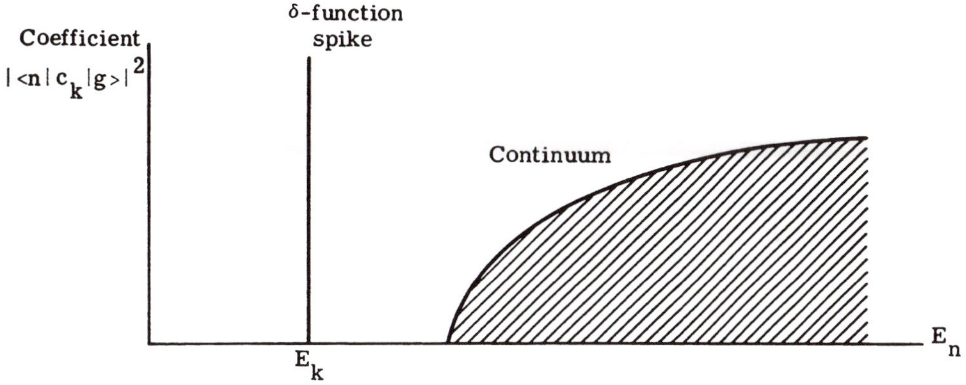

Figure 30

us infinitely more of the exact quasi-particle function than of any of the background. In terms of energy spectra, we may draw Figure 30. Thus if we multiply by an energy delta-function filter like $\int_0^\infty dT\ e^{-iET}$ we can easily pick out the quasi-particle state and its energy. This trick of obtaining exact results by studying the behavior in the $T \to \infty$ limit is a characteristic formal device of the M.B.T.

Kohn, Ambegaokar (69), and recently Blount (45) have established a considerable number of exact results about the N + 1 electron problem based on these ideas. One result is to show rigorously that in the equivalent to the one-band approximation one may use the quasi-particle energy curve E(k) as an effective kinetic-energy part of a one-band hamiltonian, just as in the theory of Bloch electrons one may treat the hamiltonian as E(k) + V(R). A second result is to show that in the long-range limit the potential is corrected by the macroscopic, static dielectric constant κ; an electron is attracted to a stationary charge q by a potential $eq/\kappa r$, if the electron is far from the charge.

This last result has received a great deal of attention in the literature. Although it is trivial from a physical point of view, the mathematical demonstration is not, and it brings out some interesting points. I cannot in this course go into the full derivation, but I think it will be instructive in the techniques of the N + 1 body problem to show roughly what is involved in it.

Suppose there is built up — say by the introduction of the external charge e — an electrostatic potential V(r) in the material, which of course gives rise to a perturbation term in the hamiltonian

$$\int dr\ \Psi^\dagger(r)\,V(r)\Psi(r) = \frac{1}{N}\sum_{q,k} c^\dagger_{k+q}\,c_k\,V(q) \tag{3}$$

where

$$V(q) = \frac{1}{\Omega}\int dr\ e^{iq\cdot r}\, V(r)$$

$$= \frac{4\pi e}{\Omega q^2}$$

for an external charge e.

We have assumed here that the ion cores do not move when the charge is introduced, so that the potential is not reduced by the dielectric constant corresponding to the ion-core motion; this is reasonable in a nonpolar semiconductor, and can be taken into account anyway in other cases.

A typical matrix element of (3) between two quasi-states K, K+Q is

$$\left(\Psi_{K}, \sum_k c^{\dagger}_{k+Q}\, c_k \Psi_{K+Q}\right) V(Q) = M_K(Q)\, V(Q)$$

where obviously all matrix elements with $q \neq Q$ must disappear by momentum conservation. The theorem of Kohn and Ambegaokar is that the scalar product in this equation is equal to $1/\kappa$, where κ is the dielectric constant of the electron gas, at least in the limit $Q \to 0$, which corresponds to very long range forces [the long-range part of V is of course $V(Q \to 0)$].

This is a rather surprising theorem at first sight because precisely in the limit $Q=0$ the matrix element is $(\Psi_K, \Sigma_k n_k \Psi_K)$ = total number of particles = N+1. The reason for this discrepancy is clear, however; the disturbance due to a single quasi-particle has some basically long-range components because the charge of that electron will polarize the entire crystal out to $r \to \infty$, and this long-range effect leads to peculiar behavior in the limit $Q \to 0$. That this is the reason can be demonstrated in a very direct way. Let us look at a sample of a perfect insulator to which we have added a single electron, not in a particular quasi-particle state K but in a wave packet of the sort to which I referred earlier, made up of a linear combination of states around a given K_o and localized near a point R_o, which we shall choose as the origin:

$$\Psi_{R_o=0,\,K_o} = \frac{\text{const.}}{\left[N(\Delta K)^3\right]^{1/2}} \sum_K e^{-\frac{1}{2}(K-K_o)^2/(\Delta K)^2}\, \Psi_K \tag{4}$$

Let us first look physically at the problem. We have added an extra electron within a region of size $\Delta R \approx 1/\Delta K$ at the center of the insulator (Figure 31). We think therefore that there is one extra electronic charge within this region. If, now, we consider the resulting polarization of the sphere of dielectric from ΔR out to the outer radius R of the sample, this sphere will be polarized by the field E of the

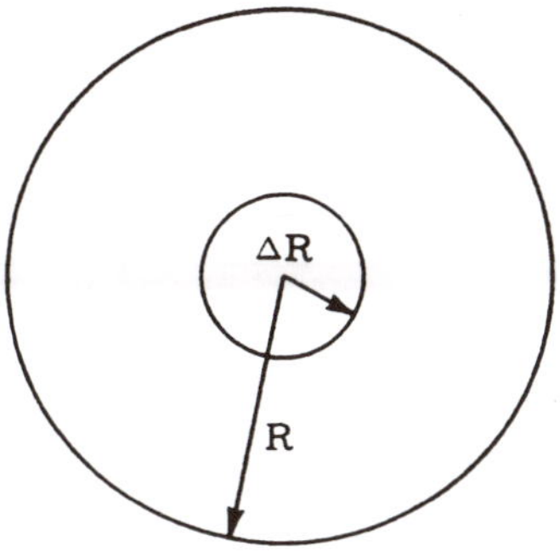

Figure 31

electronic charge, the polarization being $P = [(\kappa - 1)/4\pi]E$. This leads to a total surface charge on the inner and outer surfaces of the sphere of

$$+ (\kappa - 1)(\Delta R)^2 E(\Delta R) \quad \text{inner}$$

$$- (\kappa - 1) R^2 E(R) \quad \text{outer}$$

and E is the field caused by the net charge,

$$E(\Delta R) = \frac{e - (\kappa - 1)(\Delta R)^2 E(\Delta R)}{(\Delta R)^2}$$

or, $Er^2 = e/\kappa$, and the surface charges are $\pm [(\kappa - 1)/\kappa]\, e$.

Actually, the sphere is not cut physically by the boundary at ΔR, so that this surface charge is approximately compensated by the surface charge on the inner dielectric sphere. The actual location of the polarization charge is specifiable only roughly, as being in the region where the quasi-particle is. All we can definitely determine is that a fraction $(\kappa - 1)/\kappa$ of the electronic charge appears at the outer surface of the sphere, only a fraction $1/\kappa$ staying behind to be capable of interacting with any local externally inserted charge q.

This is a particularly simple example of a peculiar phenomenon characteristic of all problems involving charge carriers in polarizable media: that an appreciable fraction of the moving charge is always effectively at the surface of the sample. In the metal, in fact, where $\kappa(q \to 0) = \infty$, all the charge is in principle on the surface, in the dielectric only $(\kappa - 1)/\kappa$. Another way of putting it is that the infinite range, $Q = 0$, term always behaves anomalously and discontinuously. We may also interpret this as a screening effect — at least partially; the potential of interaction of the electron with other charges is reduced by a factor κ at long distances.

It remains to show the relationship of these facts to the Kohn-Ambegaokar identity involving the matrix element

$$\lim_{Q\to 0} M_K(Q) = \lim_{Q\to 0} \sum_k (\Psi_K, c^\dagger_{k+Q} c_k \Psi_{K+Q}) \tag{5}$$

Note that this, like the constant $Z^{1/2}$, is a "cross" scalar product between quasi-particle and bare-particle quantities. This is, in fact, precisely analagous to the "vertex renormalization constant" of quantum field theory, just as Z is to the "wave-function renormalization" constant Z of that theory.

We evaluate Eq. (5) by computing the charge in a sphere containing the sphere ΔR within which the quasi-particle is to be found. Introduce a smooth function f(r), which is unity for $r = \Delta R$ but falls off to zero at some radius R_1, where $R \gg R_1 \gg \Delta R$. Then the charge within the sphere of radius R_1 is given by

$$e\left(\Psi_{k_o=0,\,K_o} \int dr \; f(r)\Psi^\dagger(r)\,\Psi(r)\,\Psi\right) \tag{6}$$

Let us Fourier-analyze this expression using

$$\Psi^\dagger(r)\,\Psi(r) = \sum_{Qk} e^{iQ\cdot r} c^\dagger_{k+Q} c_k$$

and the Fourier expansion (4) of the wave packet $\Psi(R=0)$. Equation (6) becomes

$$\frac{e \times \text{constants}}{(\Delta K)^3 N} \sum_{K,K',k,Q} \left(\int dr \; f(r)\, e^{iQ\cdot r}\right)\left(\Psi_K, c^\dagger_{k+Q} c_k \Psi_{K'}\right)$$

$$\times \exp\left[-\frac{1}{2}\left(\frac{K-K_o}{\Delta K}\right)^2 - \frac{1}{2}\left(\frac{K'-K_o}{\Delta K}\right)^2\right]$$

Now,

$$\int dr \; f(r)\, e^{iQ\cdot r} = f(Q)$$

f(Q) has three properties of importance: (a) $f(Q=0) = (4\pi/3)R_1^3$, the "volume" of f(r); (b) $f(Q > 1/R_1) \simeq 0$, so that f(Q) is more steeply peaked than ΔK in Q space, and $f(\Delta K) \simeq 0$; and (c) $\Sigma_Q f(Q) = 1$. Momentum conservation requires that $K' = K+Q$, and property (b) then makes the exponential factor just

$$\exp\left[-\left(\frac{K - K_o}{\Delta K}\right)^2\right]$$

We expect, since K is very nearly equal to K_o, and the matrix element

$$M_K(Q)$$

has no reason to vary rapidly with K, that we may perform the sum over K assuming M_K constant

$$(6) = e \sum_Q f(Q)\, M(Q) \sum_k \frac{\text{const.}}{N(\Delta K)^3} \exp\left[-\left(\frac{K - K_o}{\Delta K}\right)^2\right]$$

and the sum is just the normalization sum for the packet. Thus the charge expression (6) becomes just $e\, \Sigma_Q f(Q) M(Q)$. As we have already remarked, at Q exactly zero

$$M(0) = N + 1$$

so that one contribution to the charge is just

$$\frac{4\pi R_1^{\,3}}{3}\; e\; (N + 1)$$

This contribution represents simply that part of the average charge of the entire electron gas contained within the sphere R_1, and must of course be precisely compensated by the ionic background charge. This part of the charge is easily recognized because it has nothing to do with the packet, since it increases indefinitely with R_1. In fact, one can show that it is unaffected even by removing the wave packet from within the sphere R_1. Thus it is purely a trivial background effect.

If, as is very reasonable, we set $M(Q \neq 0)$ = a constant, M, the remaining charge is

$$eM \sum_Q f(Q) = eM$$

[by property (c)] ; this is, then, independent of the radius R_1 and is the charge directly associated with the wave packet itself. This, as our physical reasoning has told us, is just e/κ. Thus we have achieved a physical, but clearly not a rigorously mathematical, proof of Kohn's identity. The actual proof can apparently only be carried through by diagrammatic perturbation theory. It is important to realize that the whole apparatus of N+1 body theory has indeed been proved rigorously within the bounds of perturbation theory only, even though in this case in particular we have every reason to believe that the perturbation theory actually converges and represents the physical reality accurately.

One result which is of some interest has been added to this by Blount. In effect, what we have been doing so far corresponds to the zero'th order one-band theory of Bloch electrons—i.e., to the effective hamiltonian theory

$$\mathcal{H} = E(k) + V(R)$$

Blount points out that one may achieve a full theory of the N+1 body problem by treating the N+1 body wave function Ψ_K as essentially

$$\Psi_k (r_1, r_2, \cdots, r_{N+1}) = \text{antisymmetrization of} \left\{ e^{ik \cdot r_1} U_o^{\,k}(r_1 \cdots r_{N+1}) \right\}$$

where U_o^k is a perfectly periodic function of the variables of the N+1 particles. U_o^k can then be treated very like the Bloch part of the usual wave function, and from it for instance one can derive, using Kohn and Ambegaokar's identity, the relationship

$$V(r) \rightarrow \frac{1}{\kappa} V(R) + \text{correction terms of order } \frac{\partial V}{\partial R} \cdot a$$

The correction terms, however, are quite different from those of the true one-electron case, as one would expect, because now V can polarize the atomic cores as well as the wave function of the added electron.

2. Effects of Phonons in the N + 1 Body Problem

Throughout all of this we have been speaking as though in the many-body system only the electrons of the valence band were free to move. This assumption is based on the Born-Oppenheimer adiabatic approximation, in which it is assumed that all motions of nuclei are infinitely slow relative to those of the electrons, because of their very great difference in mass. That approximation is very good in the finite molecular systems for which it was invented, where the electronic excitation energies and thus their frequencies are indeed very high relative to those of the ions, so the ions can be assumed stationary; but of course it must break down entirely in precisely the region we have found most interesting, namely, the region of the lowest energy excitations of a solid, when we cannot *a priori* consider that electronic energies are large or small relative to ionic ones.

The ionic motions of a solid which are of any importance here may be described in terms of quantized waves of vibration of the ion cores about their equilibrium positions (Figure 32). The quantized waves are, as you all know, called phonons.

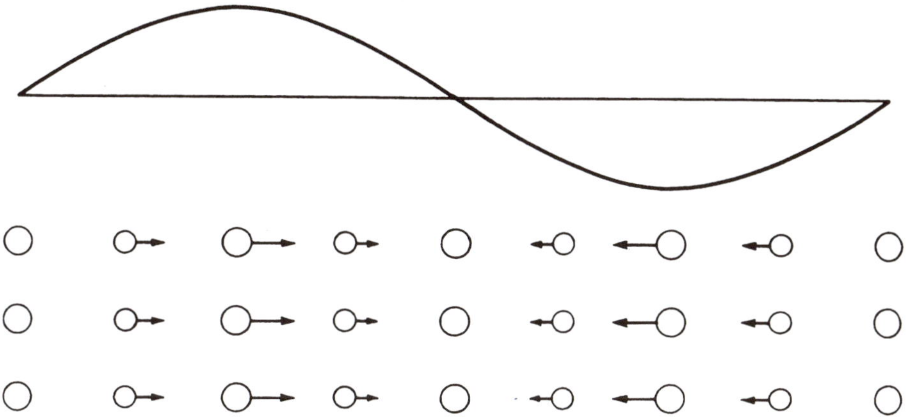

Figure 32

They are themselves elementary excitations of the solid; they were, in fact, the first set of elementary excitations to be discovered—by Einstein and by Debye, in 1910 or so. Nonetheless it may surprise you to know that many important problems connected with phonons as many-body collective excitations of a solid remain. For the time being, however, we are interested only in the phonons as modifying the properties of the quasi-particle excitations we have been discussing.

As you all know, for a solid with a single atom per Bravais lattice cell, there are three branches of the phonon spectrum, degenerating to one longitudinal and two transverse sound waves for long wavelengths and simple crystal directions; the frequencies are given for long waves by $\omega_k = c_l k$ or $c_t k$, c being the longitudinal or transverse sound velocity; the energy is $\hbar\omega_k$.

More complex solids have three or more "optical" branches of the phonon spectrum, degenerating for long wavelengths into motions of the different atoms in the Bravais cell against each other, and in cases in which the two atoms are different,

possessing a corresponding dipole moment **P** per unit volume. As we shall discuss shortly, in the optical branches the frequencies of longitudinal and transverse

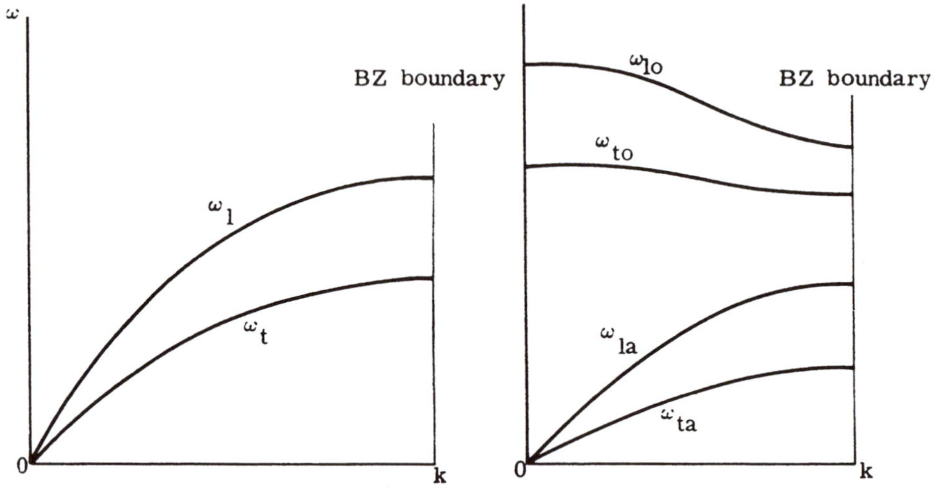

Figure 33

waves differ by a finite amount $\omega_{lo} - \omega_{to}$ at infinitely long waves. We sketch, then, phonon spectra for a monatomic and for an ionic diatomic lattice (Figure 33).

The presence of the phonons has, rather obviously, a number of consequences. First and most simply, we see that the quasi-particle state is no longer necessarily the lowest-energy excited state belonging to a given momentum or k vector; in fact usually it will be a very high energy state relative to the phonon states. This is the reason why we qualified the definition of an elementary excitation by requiring that the excited state be *of a particular type*. In the case of quasi-particles as opposed to phonons, the distinction is really quite easy, since the quasi-particle operator is a fermion and changes the number of electrons by 1, from N to N+1, while the operator $b^\dagger_K$, which creates a phonon, is approximately a boson and changes neither ion nor electron number, but in general the distinction is not so simple; for instance, it is not obvious that the state $q^\dagger_k \Psi_N$ is necessarily lower in energy than the state $q^\dagger_{k-K} b^\dagger_K \Psi_N$ with a phonon + a quasi-particle excited; both operators are fermions and both states N+1 particle states.

A much more serious problem comes in when we introduce the concept of electron-phonon interaction. This interaction has two effects: *scattering* and *polarization*. Before discussing these effects, however, I should like to say a few words about the electron-phonon interaction itself.

To get some ideas about the form this interaction may take, let us calculate it in a simple approximation. One technique, which is very simple but must be treated with caution, is the so-called "rigid-ion" model. Here we assume that the lattice potential with which the electrons interact is the sum of separate contributions caused by each ion: $V(r) = \Sigma_j V(r-R_j)$ or, in second quantization for the electrons, $V = \int dr\, \Sigma_j V(r-R_j)\, \Psi^\dagger(r)\Psi(r)$. The unperturbed wave functions which we use for the electrons are calculated as though the R_j were fixed at lattice points; the perturbation, then, must come from the displacement of the R_j, δR_j, which is caused by the presence of a phonon:

$$\mathcal{H}' = \sum_{j} \int dr\; \delta R_j \cdot \nabla V(r-R_j)\, \Psi^\dagger(r)\, \Psi(r)$$

The δR_j may be analyzed in terms of phonon amplitudes Q_K:

$$\delta R_j = \text{const} \cdot \sum_K e^{-iK\cdot R_j} Q_K$$

and the Ψ's in terms of Bloch functions:

$$\mathcal{H}' = \int dr \sum_{\substack{nn'k \\ k'Kj}} e^{-iK\cdot R_j}\, e^{ik\cdot r} u_k^{n*}(r)\, e^{-ik'\cdot r} u_{k'}^{n'}(r)$$

$$\times\, Q_K\, c_k^{n\dagger} c_{k'}^{n'} \cdot \nabla V(r-R_j)$$

$$= \sum_{kKnn'} M_{k,K}^{nn'}\, Q_K\, c_{k+K}^{n'\dagger}\, c_k^n$$

with

$$M_{kK}^{nn'} = \text{const.} \int \nabla V(r-R_j)\, \varphi_{k+K}^{n}(r)\, \varphi_k^{n'}(r)\; d\Omega_{\text{cell}} \tag{7}$$

Q_K can in turn be expressed in terms of the creation and destruction operators b_K and $b^\dagger_{-K}$ for phonons. The interaction, then, has the form of scattering of an electron with creation or destruction of a phonon (Figure 34). Much more careful

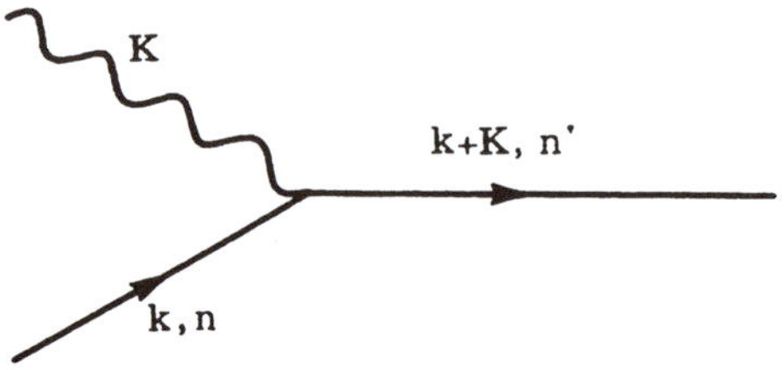

Figure 34

analyses than this rather naïve one can be carried out, and I recommend to you for a good discussion of these the appropriate chapter (Chapter V) in Ziman's book (3). Here I should like to concentrate on one limiting case, that of small k.

One finds upon looking in Ziman's book that in this limit the matrix element M comes out proportional to K, and thus the interaction is not, as would appear at first glance, proportional to the displacement $\delta R_j \propto Q_K$ of the ions, but to their strain (or "deformation"), $\nabla R_j \propto KQ_K$. This is not a terribly difficult thing to show within the one-electron theory, but it is interesting to show that it is true in general, even when many-body effects are taken into account.

This is very worth doing, because it is meaningless to discuss a displacement of the ions and ion cores alone without taking into account the fact that they carry along with them the electrons of the valence band, at least in some sense. If nothing else, we realize that, since the ion cores have a net charge, when they form a longitudinal wave of long wavelength it will build up a very large space charge in the peaks of the wave which is bound to displace the valence electron charge distribution. In general, it is obvious that the assumption that the valence electron gas as a whole does not move as the ions move is untenable. Thus a great deal of the electron-phonon interaction must be thought of as already included in the definition of the phonon excitation itself, just as the electrons which are being scattered are not really single independent electrons, but quasi-particles.

The problem is remarkably like that of renormalization in field theory; the "physical" quasi-particles and phonons we see are not the same at all as the "bare" particles we can think simply about, and their experimental properties — energy, interactions, etc. — include contributions from the cloud of disturbance surrounding the bare particles. In solid-state theory we have no divergences of the field-theoretical sort which irrevocably prevent our finding the true relationship between the "bare" properties and the "physical" properties; nonetheless, solids are such complicated things that it is often useful to resort to experiment rather than to theory in order to evaluate not only the phonon frequencies and quasi-particle energies, but their interaction. This is the idea behind the *deformation potential* theory (70). In this theory we estimate the effects of long-wavelength phonons in terms of the observable (at least in principle) effects of homogeneous strains.

Imagine then that we have, for simplicity, an insulator, which we deform elastically in some reasonably smooth fashion, so that, if the local displacements are δR_j, there is a well-defined local compression $\Delta = \nabla \cdot \delta R_j$ and, if we like, shear strain $S = \text{symm}\,\{\nabla\,\delta R_j - 1/3\,\nabla\cdot(\delta R_j)1\}$ in any local region. We allow the N-body electron gas to deform perfectly adiabatically. Then in the new, deformed lattice we can form a wave packet Ψpacket (k_o, r_o) about a momentum k_o and in the region of size $1/\Delta k$ about r_o. This region is assumed to have a reasonably uniform deformation Δ, S; otherwise, of course, the lattice would not be adequately periodic, so that we could not define the wave vector k_o to within the uncertainty Δk. This new wave packet of exact quasi-particles has a new energy $E'_n(k_o)$, which differs from the old one because the lattice is deformed locally; if there were no deformation, clearly there would be no energy change at all, no matter how much the *displacements* of the atoms might be. Since the atomic motions in a phonon are rather small, it is reasonable to expect the band energy to be linear in the deformation, and the linear constant of proportionality is called the "deformation potential" constant E:

$$E'_n(k_o) - E_n(k_o) = E_\Delta(k_o)\,\Delta + E_S(k_o)\,S + \cdots \tag{8}$$

This *empirically defined* potential is the true scattering potential for the quasi-particles $q_k^\dagger$, including the nearly infinite changes in Ψ_g caused by the deformation, as well as the modification of the quasi-particle wave function itself. This may be shown as follows. The operator which creates a wave packet around momentum k_o and position r is

$$q_{k_o}^\dagger(r) = \int dk \quad f(k - k_o)\; q_k^\dagger\; e^{ik\cdot r}$$

and the number operator which counts whether or not such a wave packet has been created is

$$n_{k_o}(r) = q_{k_o}^\dagger(r) q_{k_o}(r)$$

Then an operator which will have the same effect as the energy (8) on an actual packet wave function might be (we shall verify later that it does have the right effect):

$$\sum_{k_o} \int dr\ E_\Delta(k_o)\ \Delta(r) n_{k_o}(r)\ +\ (\text{same for S})$$

Working this out, we get

$$\sum_{k_o Kkk'} \int dr\ f(k-k_o) f(k'-k_o)\ e^{i(k-k')\cdot r}\ q_k^\dagger q_{k'}\ E_\Delta(k_o)\ e^{iK\cdot r}\ \Delta_K$$

$$= \sum_{k_o kK} f(k-k_o) f(k+K-k_o)\ q_k^\dagger q_{k+K}\ \Delta_K E_\Delta(k_o)$$

We must assume that K is smaller than the spread of the wave packet, or the f factors will not overlap correctly; this is the requirement already expressed, that the wave packets must be confined to a small enough volume so that Δ is uniform within them. If K is small enough, we can neglect it in $f(k+K-k_o)$ and then make use of the completeness relationship for wave packet functions,

$$\sum_{k_o} f^2(k-k_o) = 1$$

If then the packet spread is in turn small with respect to the range in k_o over which $E_\Delta(k_o)$ varies, we get finally

$$(8) = \sum_{kK} E_\Delta(k) q_k^\dagger q_{k+K}\ \Delta_K \tag{9}$$

This expression may be verified by applying it to the appropriate wave packet,

$$\sum_{kK} E_\Delta(k) q_k^\dagger q_{k+K}\ \Delta_K \sum_{k'} f(k'-k_o) q_{k'}^\dagger e^{ik'\cdot r} \Psi_g$$

$$= \sum_{kK} E_\Delta(k) \Delta_K\ e^{iK\cdot r} f(k+K-k_o) q_k^\dagger\ e^{ik\cdot r} \Psi_g$$

And, again, if $K \ll \Delta k$, the width of f, this just becomes

$$E_\Delta(k_o)\,\Delta(r)\,q^\dagger_{k_o}(r)\,\Psi_g$$

in agreement with Eq. (8).

In Eq. (9), then, E(k) plays the role of the matrix element $M_{k,K}$ in Eq. (7) for scattering, in the limit $K \to 0$ of very long wavelength phonons, except that $M_{k,K}$ multiplies the displacement Q instead of the deformation Δ. While it is not hard to show in a direct calculation using Hartree-Fock theory that $M_{k,K}$ is proportional to K, the argument we have given retains its validity in the presence of large many-body effects, and relates the scattering matrix element directly to experimentally observable phenomena.

It is not well known that the above argument fails completely in those crystals the symmetry of which permits a piezoelectric effect. In such crystals a deformation, especially a shear deformation Y, cannot only change the effective potential for the electrons but can set up an electric field **E** proportional to the strain. This gives us a potential term

$$E'(r) - E(r) \quad \propto \quad V(r) = e\mathbf{E}\cdot r \quad \propto \quad r\,S(r)$$

which is proportional to the total displacement, not the strain. Thus in such a case the so-called "deformation potential theorem" does not hold. Hopfield has analyzed some of the rather strange consequences which may result from that; it leads to very anomalous scattering and polarization effects for slow electrons.

In the more usual metals and semiconductors, however, the theorem does hold. This is particularly important in the case of degeneracy or near-degeneracy of two or more bands, where the theorem is valid, but it is most convenient to allow the deformation potential E to become a general matrix $E^{nn'}$ in the band indices. In this case direct computation from the rigid ion model does not give correct answers as easily.

In the presence of this electron-phonon interaction, the quasi-particles $q^\dagger_k$ may be scattered from one point to another in the Brillouin zone. At absolute zero, when there are no phonons present, the absolutely lowest quasi-particle state $q^\dagger_k\Psi_g$ in the band is still not scattered, because it cannot emit a phonon, because it is the only state of the given momentum, energy, and number of particles; but quasi-particle states of higher momentum can scatter with the emission of a phonon, as soon as $dE/dk > \hbar\,(d\omega/dk)_{phonon} = \hbar c$, where c is the lowest velocity of sound in the material (Figure 35). If the quasi-particles have an effective mass of unity and c has a typical value of 10^5 cm/sec, this is

$$\frac{\hbar^2}{m}(k - k_0) > \hbar c$$

$$k - k_0 > \frac{mc}{\hbar} \sim 10^5 \quad cm^{-1}$$

$$\frac{\hbar^2}{2m}(k - k_0)^2 \geq \frac{mc^2}{2} = \frac{1}{2}10^{-17} \text{ erg} \simeq 0.04^oK$$

That is an enormously low energy, so that for practical purposes we never deal with quasi-particles which cannot undergo scattering with emission of a phonon.

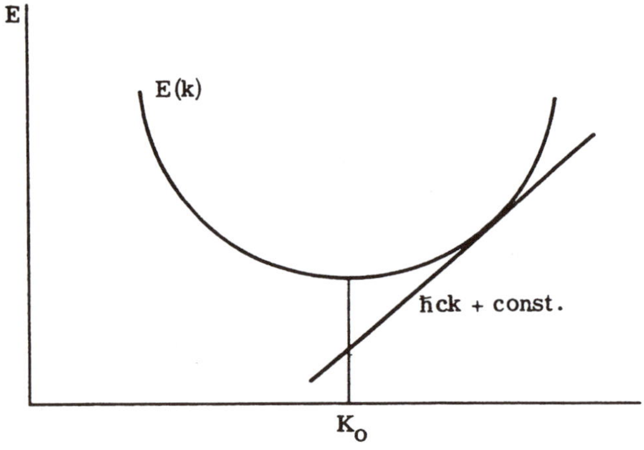

Figure 35

We realize immediately that this forces us to modify the original definition of a quasi-particle elementary excitation, because the quasi-particle state is in this case no longer a rigorous eigenstate of the entire system. This is a general phenomenon; in almost all cases the elementary excitation states are not quite exact eigenstates because of the possibility of scattering in one form or another.

As far as I know no one has ever given an absolutely exact mathematical definition of a quasi-particle in the presence of scattering, but there are a number of ways of defining quasi-particles which can be satisfactory in any given case. These may be classified under two philosophical approaches, which we will call the "pseudo-hamiltonian" approach and the "full renormalization" approach.

The first of these general philosophical approaches to this problem is essentially that which we have already carried out in our discussion of the deformation potential. That is, we try to define the quasi-particles and the lattice motions as best we can, completely independently of each other, and to leave their interaction in the problem explicitly. In the deformation-potential discussion we essentially went into the solid and put rigid clamps on all the ion cores, holding them absolutely fixed in the positions corresponding to some long-wavelength lattice deformation. From this we got a certain dependence of the quasi-particle energy on the deformation, which, when we removed the clamps, served as an electron-phonon interaction energy. In formal terms, we set up an effective hamiltonian for the quasi-particles,

$$\mathcal{H}_{\text{eff}} = \int dr \int dr' \; q^{\dagger}(r)\, H_{\text{eff}}(r, r')\, q(r')$$

where $H_{\text{eff}}(r, r')$ is a linear function of the local deformations $\Delta(r)$ and $S(r)$. We can then introduce $\Delta(r)$ and $S(r)$ as quantized coordinates of the many-body problem themselves, which will have a quadratic self-energy $\Sigma_k \hbar\omega_K^o (P_K^2 + Q_K^2)$ which must be added in. The interaction terms then can, we hope, be taken into account by perturbation theory, and lead to a scattering of the quasi-particles, as follows: The scattering hamiltonian is the deformation potential, Eq. (9):

$$\mathcal{H}' = E_{\Delta} \sum_{kK} \Delta(K)\, q^{\dagger}_{k+K} q_k$$

The normalization of the relationship between $\Delta(K)$ and a phonon creation operator b_K^* may be obtained from the energy expression:

$$\frac{S}{2} \left|\Delta\right|^2 = \text{potential energy} = \frac{1}{2}\,(\text{total energy}) = \frac{1}{2}\, \hbar\omega_K (n_K + \tfrac{1}{2})$$

where S is the compressional stiffness ρc^2 and $\omega_K = cK$ is the frequency of the phonon. Thus we may take

$$\Delta(K) = \frac{1}{2} \sqrt{\frac{\hbar K}{\rho c}}\, (b_K^* + b_{-K})$$

The expression from time-dependent perturbation theory for the inverse mean free time, or transition rate, out of a single state into a continuum of states, i.e., out of k into the states in which a phonon K has been spontaneously emitted, is

$$\frac{\hbar}{\tau} = 2\pi\, |\text{M.E.}|^2 \frac{dn}{dE}$$

M.E. means the scattering matrix element, of course. Physically the only interesting case is that in which the momentum of the electron is reasonably large, so that

$$\frac{\hbar^2 k^2}{2m} \gg \hbar k c \qquad k \gg K_o = \frac{mc}{\hbar}$$

Under these circumstances the change in momentum caused by the phonon is much greater than that in the energy, so that $|k + K| \cong |k|$ (Figure 36). Then dn/dE is just the density of states in the electronic spectrum.

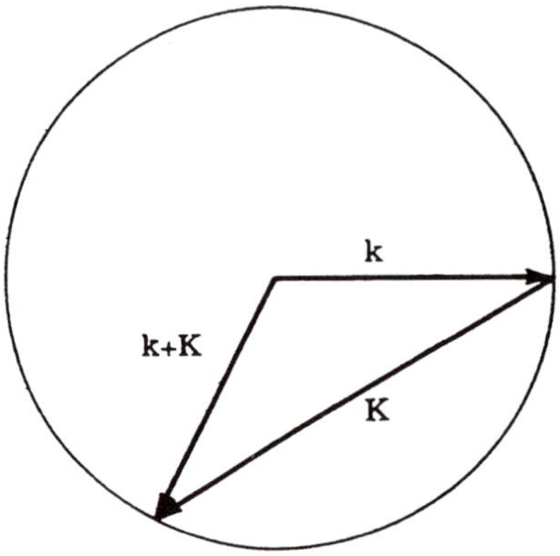

Figure 36

I shall leave it as an exercise to calculate the lifetime before scattering; the result is

$$\frac{\hbar}{\tau} = \frac{4\pi^2}{3} E^2 \left(\frac{m^2}{\hbar^3 \rho c}\right)\left(\frac{\hbar^2 k^2}{2m}\right) \tag{10}$$

If the lattice constant is a, $\rho \cong M/a^3$ and $\hbar^2/ma^2$ is of the order of an electronic energy ~ 1 Ry; $\hbar c/a$ is of the order $\sqrt{m/M} \times 1$ Ry. E, the deformation potential constant, can also be expected to be of the order 1 Ry, because a 100 per cent deformation will certainly change the electrons' energies completely; so we may say

$$\left.\frac{\hbar}{\tau}\right|_{\left(E_k = \frac{\hbar^2 k^2}{2m}\right)} \sim \sqrt{\frac{m}{M}}$$

We see that $\hbar/\tau$ is indeed quite small, and vanishes satisfactorily for the lowest-energy quasi-particles.

Associated with the scattering there is an admixture of states near the surface of constant energy: for a given energy difference ΔE, this is given in perturbation theory by

$$\frac{|M.E.|^2}{(\Delta E)^2}\ dn$$

so that the total admixture of states more than ΔE from the quasi-particle state in energy is

$$|M.E.|^2 \int_{\Delta E}^{\infty} \frac{dn}{dE}\ \frac{dE}{E^2} \cong \frac{\hbar}{\tau}\ \frac{1}{\Delta E}$$

This obviously diverges if we allow states of arbitrarily low energies to be admixed. This means that any attempt to define the quasi-particle state as an exact eigenstate in the presence of the scattering must fail. As I have emphasized, we can approach this in two ways, of which the first is the effective hamiltonian theory we have just described, in which the quasi-particles are exact eigenstates of a "clamped" system and the electron-phonon interaction is retained in the total hamiltonian with which we attempt to solve any problem of transport, etc.

The second, contrasting, method is to attempt to deal with the system as a whole, and to define quasi-particle states which decay in time because of scattering. This is the method more popular with modern many-body theorists; see Nozières (68). As we shall shortly discuss, such quasi-particle states have not only a renormalization constant Z but a complex energy $E_k + i(\hbar/\tau)$; we say that they are "complex poles of the Green's function." These relate to the propagators of the fully renormalized theory; an effective electron-phonon interaction can be introduced as the renormalized vertex part of the theory, etc. Carrying out the whole of transport theory, etc., in this fully renormalized form is as yet hardly more than a program, but no doubt it is entirely feasible. In semiconductors and insulators, certainly the first approach has so far been adequate, and it avoids the necessity for carrying about various unphysical renormalization parameters such as Z.

Nonetheless, one could argue that the first approach is a bit ambiguous and unphysical, because the real quasi-particles carry around with them an extensive cloud of polarization of the phonons, especially the shorter-wave and optical ones. But one may imagine that we put clamps only on those longer-wavelength phonons which *can* truly scatter the quasi-particles; this will hardly affect the polarization cloud of the quasi-particles but will allow us to define them quite rigorously. In particular, the quasi-particles, being related to exact eigenstates of an N+1 body system, have fixed momenta and carry the charge e except for the cautions about handling the $Q \to 0$ limit, which we have already discussed. They will obey the usual type of transport equation with the usual Lorentz force in the presence of fields.

This approach to the problem of quasi-particles in insulators seems to be perfectly rigorous and accurate, at least in nonpiezoelectric crystals and at sufficiently low temperatures. In certain cases of extremely strong electron-phonon coupling, however, the effect of the polarization cloud which accompanies the electrons as they travel may be rather large. This characteristically occurs in certain d-band oxide crystals, where the electron in a d-shell level on a given ion may polarize the surrounding ions rather severely, by a large amount relative to their zero-point fluctuations. The amplitude of the matrix element for hopping of the electron to a different d-shell ion may be written

$$b_{d_1 d_2} = \left(\Psi_{d_1} | \mathcal{H} | \Psi_{d_2}\right)$$

$$\simeq \left(a_{d_1} | \mathcal{H}_{1_{el}} | a_{d_2}\right)\left(\Psi_{\text{lattice}}(\text{electron on } d_1) \middle| \Psi_{\text{lattice}}(\text{electron on } d_2)\right)$$

a is the ordinary Wannier function for the d band, while Ψ_d is the full many-body wave function. The second factor, the lattice wave-function overlap, will be small, which greatly increases the effective mass of the quasi-particles. This is the physical picture which corresponds to the "self-trapped polaron"— it can move, but only with a greatly reduced matrix element, and greatly increased m*. At even rather low temperatures, the probability of the electrons moving by thermal activation rather than by this tunneling process may become large, which means that transport in such materials does not resemble that in ordinary semiconductors at all (71).

At higher temperatures, even in the normal case, extra scattering by thermally excited phonons becomes much greater than the spontaneous scattering by emission. The (matrix element)^2 factor in the phonon lifetime must be multiplied by an occupation factor n+1 for emission, or n for absorption, where $n = 1/[\exp(\hbar\omega/k_B T) - 1]$. This may become quite large for the relevant phonons in a semiconductor, which are those for which

$$K_{\text{phonon}} \sim k_{\text{electron}}$$

where

$$k_{\text{electron}} = \frac{\sqrt{2m^* k_B T}}{\hbar}$$

$$n(K) \sim \frac{k_B T}{\hbar c K} = \frac{v_{el}}{c} \sim 10 \quad \text{at} \quad 50 - 100^\circ K$$

The thermal scattering of the slowest electrons can become extremely severe, even at low temperatures, although this does not seem to be very important physically. At higher temperatures, our formula (10) for the scattering rate $\hbar/\tau$ is multiplied by $k_o T/\hbar c K$, and since K is of order k and $k_B T$ varies as k^2 we get

$$\frac{\hbar}{\tau} \sim \frac{E^2}{\rho c^2} k^3 \sim T^{3/2}$$

The $T^{-3/2}$ dependence of mobility $\mu = e\tau/m$ is a familiar result in semiconductors. Note that as T increases $\hbar/\tau$ will fairly rapidly become large compared to $k_B T$.

I do not know of any discussion of this case; in it the concept of quasi-particles should not be used without additional justification. It occurs in the region of mobilities around 10 at room temperatures, a not uncommon sort of value in poor semiconductors.

Effects in metals in case the electron-phonon scattering becomes strong are much less well understood than in insulators and semiconductors. In the metal at low temperatures spontaneous emission cannot take place for electrons near the Fermi surface, because the final electronic state is full. The thermal absorption and emission processes which lead to electronic scattering lifetimes have two different temperature dependences. At low temperatures only long-wavelength phonons are excited. As you will remember, the scattering matrix element is proportional to $K^{1/2}$, and the possible final states lie approximately on a sphere in K space, the surface element of which is $\propto K\ dK$. Thus

$$\frac{\hbar}{\tau} \propto \int_0^{K_F} \frac{K^2\ dK}{\exp(\hbar c K/k_B T) - 1} \propto \left(\frac{k_B T}{\hbar c}\right)^3$$

At temperatures above the Debye temperature, all phonon states have an occupation number proportional to T and $\hbar/\tau \propto T/\theta_D$ (Figure 37).

We can distinguish two cases in the high-temperature limit, $\hbar/\tau >$ or $< kT$. When $\hbar/\tau < kT$, perturbation theory of either type is probably quite satisfactory,

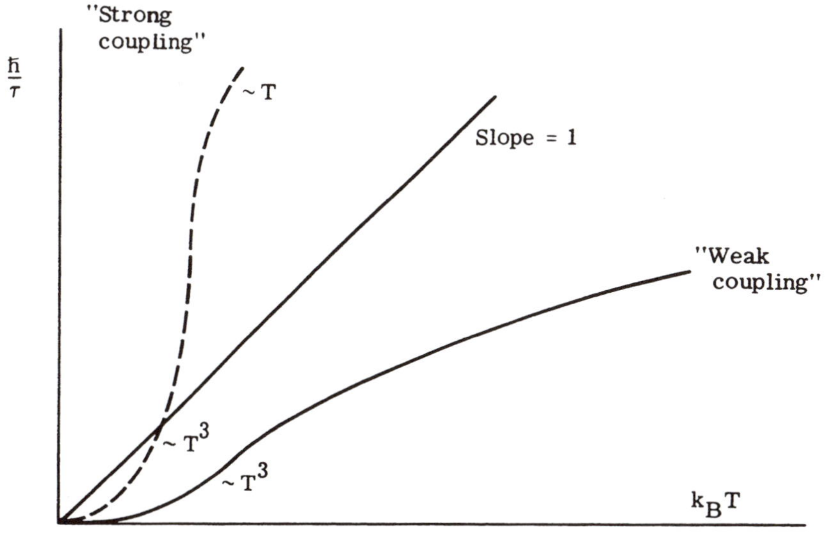

Figure 37

as can be verified both from a consideration of polarization effects at low temperatures or from the obvious comment that the quasi-particle energy breadth is small compared to the relevant energy kT at all temperatures. In the opposite, or strong-coupling, case, however, although it may be argued that transport theory can be valid (see Ziman's book) it can be shown, as in the old Fröhlich-Bardeen theory of superconductivity (72), that polarization effects will be very large.

Summarizing, we may say that the concept of quasi-particles in the presence of phonon scattering and polarization—i.e., of polarons in some generalized sense—is fairly satisfactory in semiconductors and insulators at low to moderate temperatures, and in metals such as Na which have weak electron-phonon coupling, but that at high temperatures and for strong coupling its relevance has not been demonstrated.

C. QUASI-PARTICLES IN METALS: THE FERMI LIQUID

The Landau "Fermi liquid" theory (73) may perhaps be thought of as a much more problematical application of the "clamping" or effective hamiltonian idea. Landau's theory is an attempt to apply the quasi-particle idea to the case of a degenerate Fermi gas with strong interactions, such as the electrons in metals, or liquid He_3 at low temperatures.

In this case the scattering is an effect of interactions among the electrons themselves, so that to justify the use of quasi-particles entirely in terms of a clamping theory involves putting clamps on the motion not only of the ion cores but of the electron gas itself. But this cannot be done in any very rigorous way.

Landau's approach is to apply his "clamps" in the sense of prescribing what the distribution of quasi-particles $n(k, r)$ is to be as a function of position in the metal. Then, just as we did for quasi-particles in the presence of phonons, he supposes that the energy of a given wave packet $q_k^\dagger(r)$ varies depending on the density of other quasi-particles in the same region:

$$E_k(r) = E_k^0 + \sum_{k'} f_{kk'} n(k', r)$$

It is necessary to keep the index k, indicating which k value the packets of quasi-particles are centered on, because the low-energy quasi-particles may range over a whole Fermi surface in momentum space. Thus we must handle the distribution function n with caution, because k and r are not commuting variables and must be specified only loosely.

We have not as yet stated what the quasi-particle packets are of. In this case even the lowest exact excited states of the system are enormously complex. This may be seen directly from the lowest order in perturbation theory. Consider a particle k which has been excited above the Fermi surface to k_F+a, resulting in one of the simplest excited states of the noninteracting Fermi system. This electron may be scattered to the state $k = k_F+a \rightarrow k - q$ by a process which excites a second electron k' to $k'+q$, so long as all the energies in the final state are above the Fermi energy so that the process is not forbidden by the Pauli principle

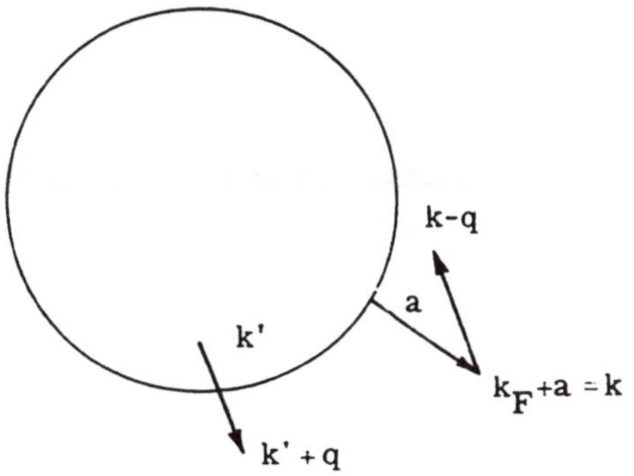

Figure 38

(Figure 38). It is no doubt well known to you that the resulting mean free time varies as the square of the energy increment,

$$\Delta E = \frac{\hbar^2}{2m}\left[(k_F + a)^2 - k_F{}^2\right] \simeq \frac{\hbar^2}{m} k_F a$$

$$\frac{\hbar}{\tau} = \text{consts.} \times \frac{|V|^2 (\Delta E)^2}{E_F{}^3}$$

This may easily be seen by observing that for any final state with the hole in state k' energy conservation requires:

$$k^2 - (k-q)^2 = (k' + q)^2 - k'^2$$

or

$$(k - k') \cdot q = q^2$$

This is the equation of a sphere whose diameter is (k - k'): thus we know that k' + q must lie on the sphere whose diameter is the vector from k to k'. When a is small, this sphere can be approximated by planes through k and k', since k', k, k-q, and k'+q must all lie very near the Fermi surface (this is true except for $k' \simeq -k$, which is a very small fraction of the available states). Now clearly, if k' lies within ($a \cos\theta$) of the surface, two types of scattering process are possible: if we pick a $q \perp (k-k')$ and small ($q > |k_F - k'| \cos\theta$, $q < a \cos\theta$), we may have $k \to k-q$, $k' \to k'+q$, or $k \to k'+q$, $k' \to k-q$ (Figure 39). In either case, the possible ranges both of k' and of q for a given θ are proportional to $a \propto \Delta E$; thus the total scattering probability at this θ is $\propto (\Delta E)^2$. I leave it as an exercise to perform the integration over θ and get the full expression for the $\hbar/\tau$ of the electron of a given ΔE.

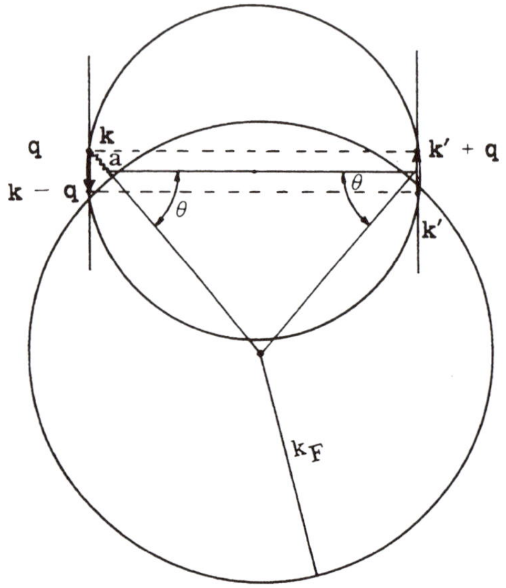

Figure 39

What gives one hope that even with scattering the quasi-particle concept has some meaning is the fact that the lifetime, at least to lowest order, of an electron becomes longer very rapidly as its energy approaches the Fermi surface. If we look at higher-order scatterings, we find that the exclusion principle reduces them even more radically than the lowest-order effect I have just discussed. Goldstone, Luttinger, Ward, and various others (74) have approached this problem from the general field-theoretical, full-renormalization approach. That is, they have shown how one can continue adiabatically, by summing perturbation theory to all orders, from the spectrum, density of states, and—what summarizes both—the Green's function of the noninteracting gas to that of one with interactions.

In the case of the insulator, you will remember that we wrote the Green's function of the system

$$G(k, t) = \langle g | c_k^\dagger(t)\, c_k(0) | g \rangle = Z e^{iE_k t} + \int dE_n\; f(E_n)\, e^{iE_n t}$$

The energy spectrum of the system could be defined in terms of the Green's function expressed in terms of frequency:

$$G(k,\omega) = i\int G(k,t)\,e^{-i\omega t}\,dt$$

$$= \frac{Z}{\omega - E_k} + \int \frac{f(E)}{\omega - E}\,dE$$

The actual energy spectrum of the states created through $c_k^\dagger$ is related to Im G(k); the existence of the quasi-particle state is signified by the appearance of a pole of G at E_k. What the many-body theory does is to show that, at least to all orders in perturbation theory, it is consistent *in the metal* to assume that G has a pole at a complex energy

$$E_k + i\frac{\hbar}{\tau}$$

so that $i\hbar/\tau$ is the imaginary part of the energy, or the breadth of the associated spectrum. The fact that this breadth approaches zero as $(\Delta E_k)^2$ turns out to be just sufficient to allow the existence of a sharp Fermi surface, since the definition of the quasi-particle states then becomes exact as they approach the Fermi surface.

In the metal the Green's function as a function of time drops off rapidly at first, just as it did in the insulator, and then oscillates at a frequency $E_k/\hbar$; but instead of oscillating indefinitely the oscillations are damped at a rate $1/\tau$ (Figure 40).

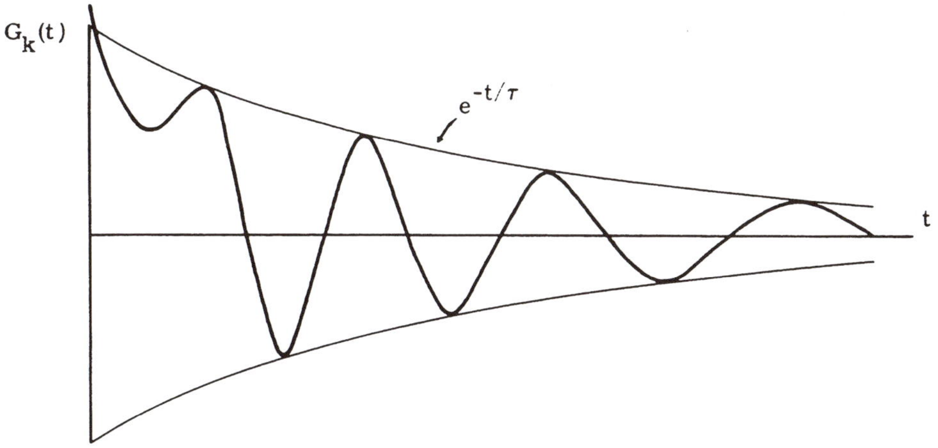

Figure 40

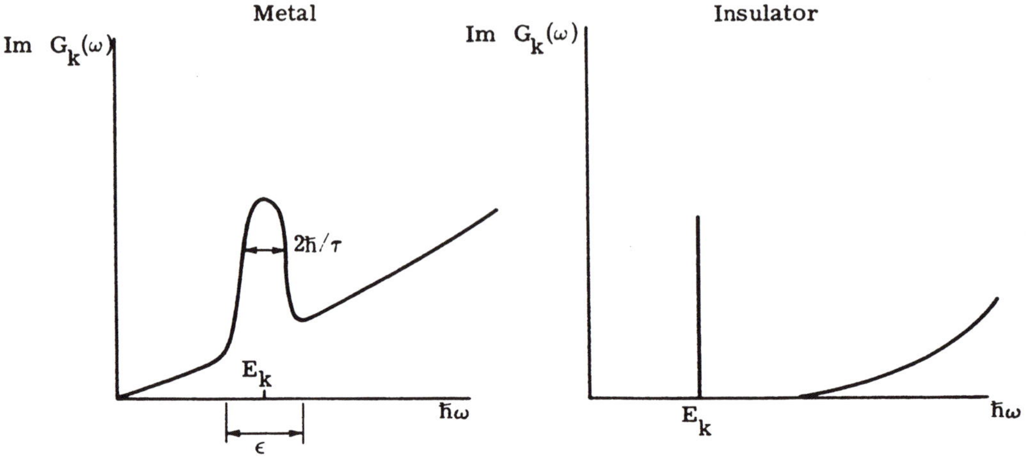

Figure 41

The corresponding spectral density is as shown in Figure 41. Again, just as in the insulator, we may reconstruct the quasi-particle state from the state $c_k^\dagger \Psi_g$ by applying an energy filter to the latter:

$$q_k^\dagger \Psi_g \propto \int_{E_k - \frac{\epsilon}{2}}^{E_k + \frac{\epsilon}{2}} c_k^\dagger \Psi_g(E)\, dE \qquad \epsilon \geq \frac{\hbar}{\tau}$$

Of course, formally this is easily expressed as a time integral:

$$q_k^\dagger \Psi_g = \int_0^\infty dt\; e^{\left(i \frac{E_k}{\hbar} - \frac{\epsilon}{2\hbar}\right)t}\, c_k^\dagger(t)\, \Psi_g$$

The quantity ϵ may be allowed to approach zero faster than E_k as $E_k \to 0$; this is the reason one can define a sharp Fermi surface and assign a definite occupation probability to a given quasi-particle state.

The limitations on the quasi-particle picture are obviously $\hbar/\tau < E_k \sim k_B T$, and *a fortiori*, $\hbar/\tau < E_F$; $\hbar/E_F$ is the time constant of the initial drop in $G(t)$.

From this kind of theory and with the aid of various identities proved by means of diagrams, Luttinger and Nozières have recently obtained directly from perturbation theory many of the results of Landau's Fermi liquid theory (75). It is apparently not yet possible to reformulate the "full renormalization" kind of theory into an "effective hamiltonian" or clamping theory. Since the "effective hamiltonian" approach is more convenient, especially for transport problems, this is a disadvantage relative to Landau's intuitive approach. In addition, the perturbation-theoretic approach must assume convergence, and it is almost an adage that P.T. never converges; this one can be shown not to.

It is a good question whether a more straightforward "clamping" approach could not be applied to the problem. We might ask whether, as in the case of the phonon theory, we could not clamp the necessary degrees of freedom of the N-particle system, without seriously altering the properties of the quasi-particles. One way of doing this might be to impose on the N-particle system a requirement that the long-wavelength degrees of freedom not be permitted to be excited—i.e., that the density fluctuation

$$\rho^q = \sum_k c^\dagger_{k+q} c_k$$

or even the individual such density fluctuations

$$\sum_{\substack{\text{some small}\\ \text{region of } k}} c^\dagger_{k+q} c_k$$

be required to be identically zero for some sufficiently small q. Since in the N+1 particle problem we could assume that for long-range effects the quasi-particle and true particle densities were the same (still with of course the very special comments about $q=0$ limits), this would be equivalent to fixing Landau's $n(k,r)$.

Unfortunately, apparently even this will not have the effect of making the quasi-particles exact excitations of the system, because of the existence of the "exchange" scattering processes $k \to k'+q$, $k' \to k-q$, which we mentioned, one for each of the long-wave processes—and, also, the interesting region $k' \simeq -k$, where all q's are allowed. Perhaps some way of getting around the exchange-scattering question may be found, but it is suggestive that the known failures of perturbation theory are connected with the $k' \simeq -k$ region (76).

It is disappointing that a full justification of the quasi-particle idea and of an effective interaction for the Fermi gas has not as yet appeared. Since these concepts are used extensively in places where the Landau quasi-classical distribution function is not a satisfactory approach—notably, superconductivity and conditions of strong impurity scattering—it would be highly desirable to have a fuller justification.

In conclusion, I should like to say that the idea of the quasi-particle is an immensely important one for solid-state physics, as of course it is for high-energy physics. The various ways of thinking rigorously, or more or less rigorously, about quasi-particles give us very valuable confidence in the bases of our physical thinking about solids. Nonetheless, one may overemphasize the rigor with which it is possible, or even useful, to treat them. As Landau says in the preface to the old version of "Statistical Physics": "We are talking here about theoretical physics, and therefore of course mathematical rigor is irrelevant and impossible." This is not quite true, but it is very close to it.

It is well to remember, first, that almost always the mathematical model we work with is a highly simplified idealization of the actual physical system already; and, second, that all quasi-particle physics is based on perturbation series in the very nature of its being, since it is an attempt to approximate a full theory by a theory resembling that of only weakly interacting particles. But we must always guess what the state from which to start perturbation theory is — that is, what the physical nature of the system is, whether metallic, magnetic, insulating, liquid, or solid. This guess may be right, but it may be terribly wrong, in which case very often the only sign may be a mild, hardly noticeable divergence of our perturbation theory. This is what happened in superconductivity. A third caution is that almost all perturbation series do fail in at least some way, because of the presence of collective excitations, which often are only to be derived from perturbation theory as sums of divergent subseries of terms. In general, I believe that the attitude is at least justifiable that the instinctive use of quasi-particles by the founders of solid-state physics and by Landau's school is hardly less rigorous than the sophisticated many-body theory, and perhaps it is more foolproof, because it has less *appearance* of being rigorous.

D. COLLECTIVE EXCITATIONS

1. Excitons

Excitons in insulators and spin waves in ferromagnets are two of the simplest type of collective excitation. They are remarkably similar in many ways, particularly so in the formal mathematics associated with them; they also have the deeper similarity that they have relatively little zero-point motion and few of the complications which that entails. In each case the reason is that, as in the N+1 body problem, one can start out by formulating the problem in terms of a nearly exactly definable ground state. In the ferromagnet, this is so because the ferromagnetic ground state is characterized (if it exists) by having the total Z-component of spin $M_Z = N/2$. All the spin-wave states, while they may be very low in energy, are characterized by $M_Z = (N/2) - 1$, and at worst only very small matrix elements of the hamiltonian connect them with the ground state.

The situation is more complex for excitons. These are, as you know, optical excitations which we can discuss either according to the Frenkel (77) or the Wannier (78) models. The Frenkel model is appropriate for tight-binding situations, such as molecular or ionic crystals, and consists of an excited atomic state moving through the crystal. For instance, it may be a state in NaCl in which one of the p electrons of the Na^+ core is excited to an s level, and the resulting excitation can move. The Wannier model is more frequently observed and perhaps more interesting; here one supposes that an electron is excited from one band to the next, but the resulting quasi-particle and quasi-hole are attracted according to the weak potential

$e^2/\kappa R$ and travel through the crystal together. In either case it is usually pretty clear how to define a fixed ground state with no excitons whatever present; for optically active excitons, for instance, such as sp transitions, the exciton state of $k=0$ has opposite parity to the ground state, so cannot mix with it. However, we shall see that this simplification is not entirely true for all excitons.

My sources here will be primarily an excellent paper by J. J. Hopfield (79), and another by E. Dresselhaus (80). I don't believe there are any good reviews; the older theory is given in Seitz.

For the purposes of qualitative physical ideas, and for comparison with spin waves, the Frenkel model is the more useful. Let us start with it and then perturb from it.

The Frenkel model envisages, at least at first, a crystal like solid Ne or Ar, in which the atoms are quite far apart and almost nonoverlapping. The crystal is an insulator, of course. We abandon for the nonce any discussion of the basic many-body problem, and treat the potential by Hartree-Fock; we shall treat the relevant interactions later by many-body theory, but it is simplest to start from Hartree, with the usual proviso that because we work with elementary excitations the many-body effects are not too bad. The last band filled we assume to be an s band. Call the Wannier function of this band $a_s\,(r-R_j)$. Because this band is extremely narrow, the off-diagonal matrix elements

$$b^s_{jk} = \int a^*_s(R_j-r)\;\mathcal{H}_{1\,el}(r)\;a_s(r-R_k)\;dr$$

are very small — with respect to what, we shall see shortly.

There is also a "conduction" band — a band of excited states, which we shall for simplicity again assume to have p-like Wannier functions

$$a_p^{x,y,z}\,(r-R_j)$$

and very small band widths (i.e., transfer integrals)

$$b^{xx'}_{jk},\quad \text{etc.}$$

The free atoms would have a certain excitation energy

$$E_o = (E_p - E_s)$$

from the s to the p state, which in such a solid would not be too seriously modified by the transformation to Wannier functions. This excitation energy, it is important to realize, is not close in magnitude to the Hartree-Fock mean energy difference between the s and p bands.

The reason is that the band energy difference is for states in which the electron has been excited completely away from its own atom, so that it finds itself on an atom which has both s and p electrons, and has therefore the extra coulombic repulsive energy

$$U = \int |a_p(r-R_j)|^2 \frac{e^2}{|r-r'|} |a_s(r'-R_j)|^2 dr\ dr'$$

in addition to the one-electron excitation energy E_o (Figure 42). Another way to

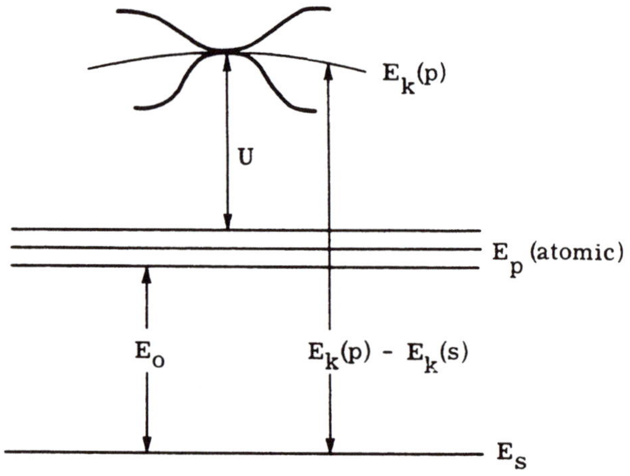

Figure 42

put it is that the electron and the hole it left behind it attract each other strongly, by the full Coulomb interaction (modified only slightly by the dielectric constant of the medium), so that it is much easier for the electron to be excited to the p state on the same atom than on a nearby atom.

Now it is clear that if the b's are small compared to U, then the Frenkel theory will be pretty well correct, because neither the hole nor the electron will tend to hop away from their partners. Then the exciton

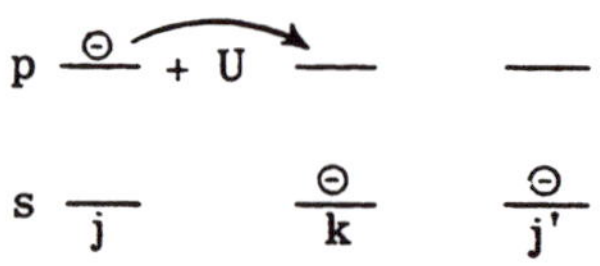

will be tightly bound, and we shall have a large gap between the continuum of s hole - p electron band states and the excitons. In the other case, $U < b$, it will be relatively easy for the electron and hole to part, and the Wannier picture will be much better. We shall discuss that case later.

On the other hand, even when the coupling of electron to hole is tight, the pair can move through the crystal together. The most important mechanism for this in the present case is the interaction of the electrons with those on nieghboring atoms. This interaction may be written

$$\int \Psi^\dagger(r)\Psi^\dagger(r')\frac{e^2}{|r-r'|}\Psi(r')\Psi(r)\,dr\,dr' \tag{11}$$

where Ψ may be expanded in terms of creation and destruction operators for the Wannier functions, c_j^s and $c_j^{p_x}$ $c_j^{p_y}$ $c_j^{p_z}$:

$$\Psi^\dagger(r) = \sum_j \sum_{n=s,p_x,p_y,p_z} c_j^{n\dagger} a_n^*(r-R_j)$$

There are various terms which may have the effect of exciting one atom and de-exciting its neighbor; the most important are the "direct" or dipolar terms, of which a typical member is

$$\left(c_j^{p_x}\right)^\dagger c_k^{s\dagger} c_k^{p_x} c_j^s \int a_{p_x}^*(r-R_j)a_s(r-R_j)\frac{e^2}{|r-r'|}a_s^*(r'-R_k)a_{p_x}(r'-R_k)\,dr\,dr'$$

Integrals like this one are most easily evaluated in terms of the multipolar expansion

$$\frac{e^2}{|r-r'|} = e^2\left[\frac{1}{R_j-R_k} + \left\{(r-R_j)+(r'-R_k)\right\}\cdot\nabla\left(\frac{1}{r}\right)\Big|_{R_j-R_k} + (r-R_j)(r-R_k)\cdot\nabla\nabla\left(\frac{1}{r}\right)\Big|_{R_j-R_k} + \cdots\right] \tag{12}$$

The first nonvanishing term is the dipolar one, the last one above, which gives for our typical matrix element

$$|X_{sp}|^2 \cdot T(r_j-R_k)_{xx}$$

where X_{sp} is the dipole-moment matrix of the atomic states:

$$X_{sp} = e\int a_{p_x}^*(r)\,x a_s(r)\,dr$$

and T is the dipolar interaction operator

$$T_{jk} = \frac{1}{R_{jk}^3} \left\{ 1 - 3 \frac{R_{jk} R_{jk}}{R_{jk}^2} \right\}$$

In this dipolar interaction operator, which represents the direct electrostatic interaction of the moving electric dipoles associated with each of the atoms, those elements of the form which I have given connect states each of which has energy about E_0 above the ground state, as I sketched before. There are terms which are of the form $c_j^{p\dagger} c_j^s c_k^{p\dagger} c_k^s$, which create or destroy two excitations simultaneously; these are of some importance and we shall soon get around to discussing them, but for the time being they are clearly not as effective as the others because they have an energy denominator ΔE of $2E_0$. We have now in our hamiltonian

$$\mathcal{H}_{\text{unpert}} = \sum_{j, p_i} \frac{E_0}{2} \left[\left(c_j^{p_i} \right)^\dagger c_j^{p_i} - c_j^{s*} c_j^s \right]$$

$$+ \frac{1}{2} \sum_{j \neq k} \sum_{p_i p_{i'}} \left(c_j^{p_i} \right)^\dagger c_k^{s\dagger} c_k^{p_{i'}} c_j^s \left| X_{sp} \right|^2 \left(T_{jk} \right)_{ii'} \tag{13}$$

It is now useful to take a step which is common to almost every problem in collective excitation theory: the transformation from pairs of fermion operators to bosons. The most familiar such transformation is the special case of electronic spin operators. Suppose that instead of the three p functions there were a single unique one ϕ_p. Then the operators involved in the hamiltonian, Eq. (13), associated with the atom j would be the three combinations

$$c_{j1}^\dagger c_{j1} - c_{j0}^\dagger c_{j0} \qquad 0 = s$$

$$c_{j1}^\dagger c_{j0} \quad \text{and} \quad c_{j0}^\dagger c_{j1} \qquad 1 = p$$

In addition, we are only interested in states in which atom j is either excited or not excited — that is, in which $n_p + n_s = 1$.

These pairs of operators are precisely equivalent to a set of Pauli spin operators

$$\sigma_{jz} = \begin{pmatrix} 1 & 0 \\ 0 & -1 \end{pmatrix} \begin{matrix} p_{occ} \\ s_{occ} \end{matrix} \qquad \sigma_j^+ = \begin{pmatrix} 0 & 1 \\ 0 & 0 \end{pmatrix} \qquad \sigma_j^- = \begin{pmatrix} 0 & 0 \\ 1 & 0 \end{pmatrix}$$

acting on a pseudo-spin wave function describing the state of the jth atom.

In the first place, pairs of Fermi operators referring to different atoms commute with each other, as the spin operators do; and second, within the subspace of the wave functions referring to a single atom these spin operators have exactly the same algebraic properties as

$$c_1^\dagger c_1 - c_0^\dagger c_0 \; (\sigma_z) \qquad c_1^\dagger c_0 \, (\sigma^+) \, ; \; c_0^\dagger c_1 \, (\sigma^-)$$

Thus in that case it would be precisely accurate to write the hamiltonian in the form

$$\mathcal{H} = \frac{E_0}{2} \sum_j \sigma_{jz} + \frac{1}{4} \sum_{jk} X_j T_{jk} X_k (\sigma_j^+ \sigma_k^- + \sigma_j^- \sigma_k^+)$$

Exactly this correspondence will be used in the magnetic case.

The transformation from spin operators to bosons was discussed exhaustively by Dyson (81), and can be simplified by a trick the source of which I have forgotten. So long as two spin operators or two pairs of fermion operators refer to different atoms, they commute, just as bosons do. But when two spin operators refer to the same atom, their commutation relations are much more Fermi-like, and this is the main difficulty in setting up a corresponding set of bosons.

In the case of excitons, and in many cases of spins, this is not a very serious limitation because the total number of excitations present is rather small, and therefore the possibility of their coinciding is negligible. In such a case we might guess that it would be satisfactory to identify the spinors directly with bosons according to the obvious prescription:

$$\sigma_j^\dagger \to b_j^\dagger \qquad \sigma_j \to b_j \qquad \frac{\sigma_{jz} + 1}{2} \to n_j = b_j^\dagger b_j \tag{14}$$

The sets of operators (14) have identical algebraic properties as far as the most important pair of levels, $n_j = 0$ and 1, are concerned. But the bosons b have matrix elements connecting to an entire ladder of extra states, $n_j = 2, 3, \ldots$.

The trick for making the correspondence (14) exact is simply to cut off the communication between the bottom two members of the ladder of states and the remainder of the ladder. A "barrier" is inserted which projects out just the matrix elements connecting $n_j = 1$ and $n_j = 2$, and does not alter the $n_j = 0$ and 1 matrix elements otherwise (Figure 43). The "barrier" consists in multiplying by $(1-n_j)$ before using $b_j^\dagger$ at all points, and *after* using b_j. That is,

$$\sigma_j^\dagger = b_j^\dagger (1-n_j) \qquad \sigma_j = (1-n_j) b_j \qquad \sigma_{jz} = 2n_j - 1$$

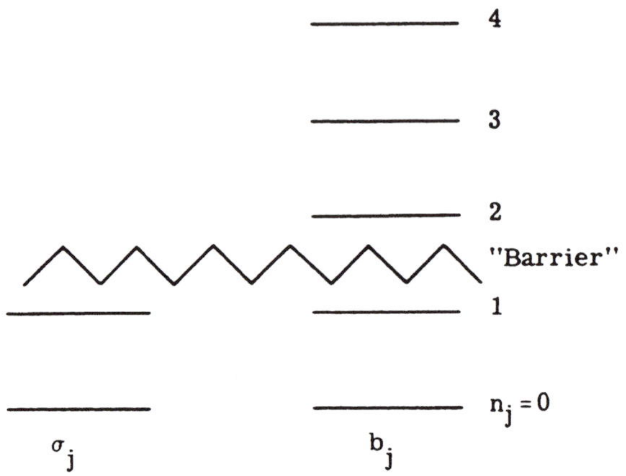

Figure 43

It is immediately verified that when $n_j = 1$, $\sigma_j^\dagger$ cannot raise it to $n_j = 2$, and vice versa.

Other techniques may be as good, as for instance the Holstein-Primakoff expansion or Wentzel's effective hamiltonian approach (82); and certain Green's function schemes are more general (93); but this is by far the simplest demonstration of the Fermion pair to boson transformation. Using this transformation, the hamiltonian in the one-excited-level case is now, still exactly,

$$\mathcal{H} = E_o \sum_j n_j + \sum_{jk} b_j^\dagger (1-n_j) T_{jk} (1-n_k) b_k \frac{|X|^2}{2}$$

The generalization of this to the case of degenerate excited states is trivial. In the case we are discussing, we can replace the pair of levels, one triply degenerate, by the bottom two levels of a three-dimensional harmonic oscillator

$$b_j^\dagger = b_j^{x\dagger},\ b_j^{y\dagger},\ b_j^{z\dagger}$$

Then in order to impose our barrier we simply multiply b^x by $(1-n_j)$, where $n_j = n_{jx} + n_{jy} + n_{jz}$; then $c_j^{p_x\,\dagger} c_j^s = b_j^{x\,\dagger}(1 - n_j)$, etc. The resulting hamiltonian in terms of bosons is

$$\mathcal{H} = E_o \sum_j n_j + \frac{|X|^2}{2} \sum_{j \neq k} b_j^\dagger (1-n_j) \cdot T_{jk} \cdot (1-n_k) b_k \qquad (15)$$

Again, (15) is perfectly exact; the "kinetic restrictions" have been replaced by a nonlinear interaction.

There are two ways to proceed from here. The second, direct diagonalization, we shall illustrate later; the first is to compute directly the equation of motion of the boson operator:

$$i\hbar \frac{db_j^\dagger}{dt} = \left[\mathcal{H}, b_j^\dagger\right] = E_o b_j^\dagger + |X|^2 \sum_{k \neq j} b_k^\dagger \cdot T_{jk} (1-n_j)(1-n_k)$$

$$- \frac{|X|^2}{2} \sum_{k \neq j} \left[(b_j^\dagger)^2 \cdot T_{jk} \cdot (1-n_k) b_k + b_k^\dagger (1-n_k) \cdot T_{kj} n_j\right]$$

We have used $[b, b^\dagger] = 1$.

If this equation is allowed to operate on the ground state, which has no excitons excited, the last two terms are identically zero, as are the correction terms in the second. It is then a linear equation for the b_j's, and may be diagonalized by the linear transformation

$$b_j = \frac{1}{\sqrt{N}} \sum_k e^{ik \cdot R_j} \beta_k$$

from atomic excitons to running waves. The resulting equation is

$$i\hbar \frac{d\beta_k}{dt} = E_o \beta_k + X^2 T \cdot \beta_k$$

where T is the Fourier transform of the dipolar interaction

$$T_k = \sum_{j \neq 0} e^{ik \cdot R_j} T_{0j}$$

$$= \sum_{j \neq 0} \frac{e^{ik \cdot R_j}}{R_j^3} \left\{ 1 - 3\hat{R}_{0j}\hat{R}_{0j} \right\}$$

Note that $\Sigma_k T_k = 0$ because of the absence of the $j = 0$ term. For very long wavelength excitons, which are the only ones of much interest, and in cubic crystals or along axes of symmetry, T may usually be diagonalized by separating the longitudinal and transverse waves; i.e., one of its principal axes is along k, two perpendicular to it.

In the limit $k \to 0$, T becomes the famous dipolar sum which is of basic importance in the theory of local fields. As $k \to 0$,

$$(T_l)_k - (T_t)_k \to 4\pi N$$

and for cubically symmetric sites j,

$$(T_l)_k \to \frac{8\pi N}{3} \qquad (T_t)_k \to -\frac{4\pi N}{3}$$

$4\pi/3$ is the appropriate Lorentz local field factor. The difference between longitudinal and transverse waves as $k \to 0$ reflects the long-range character of dipolar forces; a longitudinal dipolar wave leads to accumulations of charge at the modes and antinodes of the wave, a transverse wave does not (Figure 44).

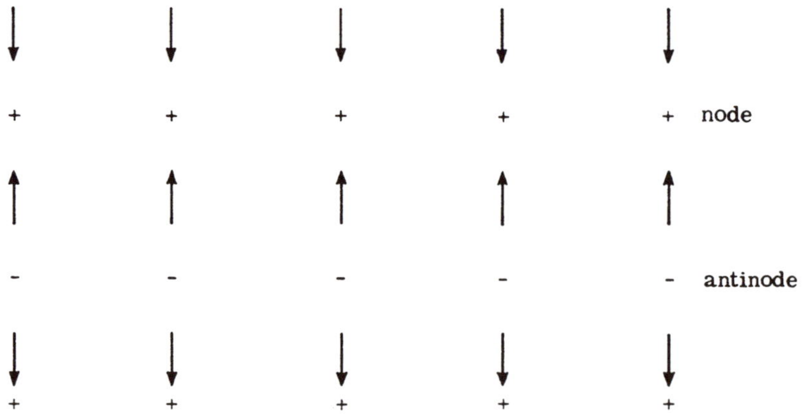

Figure 44

Thus the energies of the excitons in this approximation are

$$E_l = E_o + \frac{8\pi}{3} NX^2 + \text{const.}\ X^2 \frac{k^2}{a} + \cdots$$

$$E_t = E_o - \frac{4\pi}{3} NX^2 + \text{const.}\ X^2 \frac{k^2}{a} + \cdots$$

Probably the best treatment of dipolar sums is in a paper of Cohen and Keffer (84).

The actual values of T_l and T_t depend on details of the structure. The difference, and indeed the difference between the longitudinal and transverse exciton frequencies, even in cases in which the Frenkel model is not valid, does not so depend and is a purely long-range, macroscopic effect. As you can see, T_t is the sum of the fields due to parallel dipoles with the surface or volume charge at very large distances effectively canceled out:

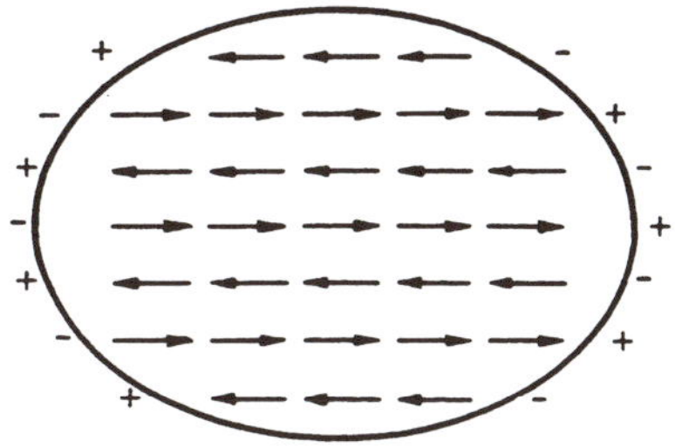

which is equivalent to

or

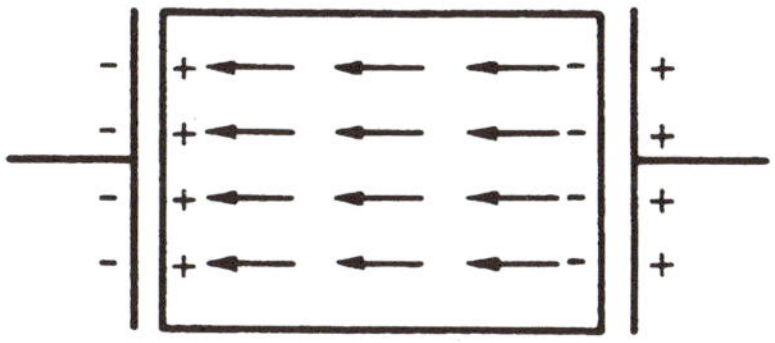

while T_l is the opposite case, equivalent to

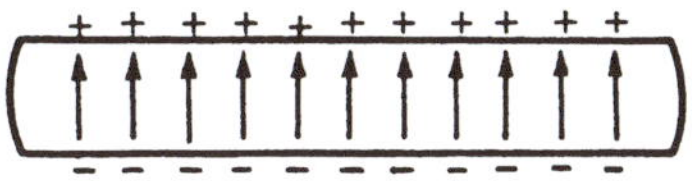

as shown above. The difference is entirely caused by the macroscopic surface or volume charges, which can be calculated by macroscopic dielectric theory. There is a complete analogy with the theory of Lyddane, Sachs and Teller (85), and others of the longitudinal-transverse frequency difference of optical phonons.

To show the direct relationship of all this to classical dielectric theory let us write down the equation of motion of a classical dielectric, which is

$$\frac{d^2P}{dt^2} + \omega_o^2 P = \beta E_{eff}$$

Here β is a coupling constant which can be evaluated in terms of dipolar matrix elements by considering the zero-frequency limit, in which we know from straightforward perturbation theory that the polarizability of the system of atoms is

$$\alpha = \frac{2NX^2}{E_o} = \frac{2NX^2}{\hbar\omega_o} = \frac{\beta}{\omega_o^2}$$

In the local-field approximation, the effective field which the atoms see is $E_{eff} = E + LP$, where L is the "local-field constant," $4\pi/3$ for cubic lattices and transverse waves. The resonant frequency is determined, then, by

$$\left(-\omega_r^2 + \omega_o^2\right) P = \beta LP$$

and assuming $\omega_r \simeq \omega_o$, this gives

$$2\,(\omega_r - \omega_o)\,\omega_o = -\beta L$$

$$\hbar\omega_r \simeq \hbar\omega_o - \frac{\hbar\beta}{2\omega_o} L = \hbar\omega_o - NX^2 L$$

This is the same expression we obtained by direct quantum mechanical means. We shall see later how to restore agreement with the more complete classical formula,

$$\frac{\omega^2 - \omega_o^2}{\omega_o^2} = -\alpha(\omega=0)\cdot L \qquad (15')$$

Hopfield's paper contains an interesting treatment of the whole exciton problem in terms of quantization of a classical dielectric medium, which I shall not repeat because we shall discuss a similar treatment of spin waves later.

I should, however, discuss Hopfield's results on the interaction of excitons with radiation, because they illustrate some interesting physical points. So far we have included in our calculations only the longitudinal part, leading to Coulomb interactions, of the electromagnetic field. The transverse radiation field does not interact with purely longitudinal excitons, but of course it is coupled directly to the transverse ones by a term of the form

$$\sum_{k,\lambda} (M.E.)_k \left(-b^\dagger_{k\lambda} a_{k\lambda} + b_{k\lambda} a^\dagger_{k\lambda}\right)$$

where $a_{k\lambda}$ are the photon creation and destruction operators, and the matrix element may be shown to be

$$i\left(\frac{2\pi\hbar c}{|k|}\right)^{1/2} \frac{\omega_o}{c} X$$

λ is a polarization index. Hopfield points out that in the ideal crystal we must not think of the coupling as leading to absorption of photons by the excitons, but as a coupling between two independent boson excitations, the spectra of which are as shown in Figure 45. The exciton velocity is enormously smaller than that of the

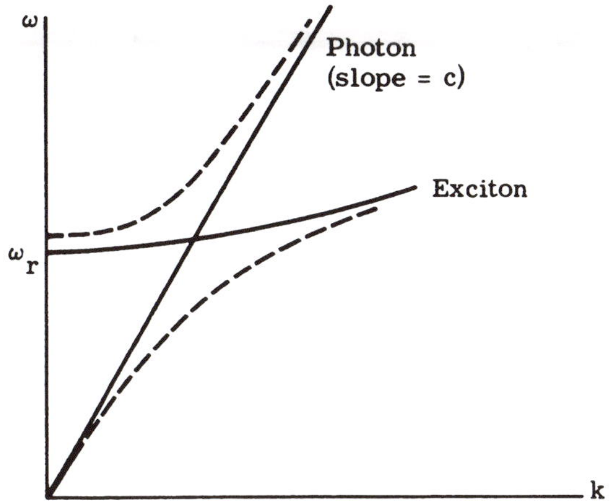

Figure 45

photon, so that the two frequencies cross at very small k. Naturally, when we introduce the coupling, the result is to split the spectra near the crossing point as shown by the dotted lines. Thus in the absence of phonons and imperfections, the photons are not absorbed but strongly dispersed at the exciton frequency. There is a region of complete opacity, or stop band, for radiation near ω_r in energy; at the sides of this band the photon velocity is greatly slowed. This is the solid-state equivalent of the phenomenon of resonance trapping of radiation; it can be thought of as the slowing down or stopping of the photon by repeated absorption and re-emission. Of course in an actual crystal it turns out that the exciton is much more strongly absorbed by such processes as multiphonon emission than the photon, so that an absorption-like process takes place near the exciton frequency. This however must be considered as a direct absorption from the mixed exciton-photon mode into phonon or impurity degrees of freedom, not as absorption of photons by the excitons themselves.

Hopfield and Thomas have succeeded in demonstrating experimentally several of these properties of excitons in CdS crystals. By observing the absorption in directions other than the principal axes they have been able to cause a slight admixture of longitudinal and transverse excitons and thus demonstrate the frequency difference of these two types of modes experimentally. In other arrangements, with the aid of magnetic fields, they have shown that the excitons interacting strongly with photons are indeed propagating with a finite, nonzero momentum rather than being stationary excited atoms, and have roughly measured the exciton velocities experimentally (86).

Another aspect of the Frenkel excitons is the role played by the high-energy dipolar terms which we have so far neglected in this treatment. The electrostatic interaction (11) also contains dipolar matrix elements like

$$\left(c_j^{p_x}\right)^\dagger \left(c_k^{p_x}\right)^\dagger c_k^s c_j^s \left(\int a_{p_x}^* \, exa_s\right)^2 \left(T_{jk}\right)_{xx}$$

which have the effect of simultaneously exciting atoms j and k. The resulting terms in the hamiltonian, written in terms of the bosons b, are

$$\frac{X^2}{2} \sum_{j \neq k} b_j^\dagger (1-n_j) \cdot T_{jk} \cdot b_k^\dagger (1-n_k)$$

$$+ \frac{X^{*2}}{2} \sum_{i \neq k} (1-n_j) b_j \cdot T_{jk} \cdot (1-n_k) b_k \qquad (16)$$

We can make X real in most cases.

The most immediately obvious thing to do is to treat these terms by perturbation theory assuming, quite justifiably, that the average energy of the appropriate excited states is just $2E_0$. We get a lowering of the total energy of the solid by

$$\Delta E_2 = -\frac{N|X|^4}{2E_0} \sum_j T_{oj} : T_{oj}$$

which is proportional to R_{oj}^{-6}, and is clearly just to be identified with the component of the Van der Waals' attraction of the atoms coming from this particular electronic transition, computed as it would be for independent atoms.

It is interesting to do this in another way. We saw that in considering the *first* exciton with the simpler hamiltonian (15) the nonlinear terms in the equation of motion of the exciton operator were exactly zero. Thus to that approximation the first exciton is an *exact* excited state of the system. We now make the typical "elementary excitation" assumption, that even with physically significant numbers of elementary excitations present, the interactions of the elementary excitations can be ignored to some degree of approximation; that is, both in (15) and in the terms (16) we are now considering, we replace $(1-n_j)$ by 1, retaining only the terms quadratic in the boson amplitudes.

One must realize that once we have included (16) in the hamiltonian, even the ground state contains a small amplitude of atomic excited states because of the pairs of excitons which are excited and de-excited. The percentage is not in fact infinitesimal, since $X^2T_{jk} \sim 1/2$ to 1 ev in good rare gases and ionic crystals, so the amplitude is of order 0.1. Thus it is a real assumption to set $n_j = 0$, but in this case a reasonably accurate one.

Let us then express the new, complete hamiltonian in this linearized approximation in terms of the boson operators $\beta_{k\lambda}$ (λ refers to the polarization of the exciton):

$$\mathcal{H} = \sum_{k\lambda} \left(E_o + X^2 T_\lambda(k)\right) \beta^\dagger_{k\lambda} \beta_{k\lambda} + \sum_{k\lambda} X^2 T_\lambda(k) \left(\beta^\dagger_{k\lambda} \beta^\dagger_{-k\lambda} + \beta_{k\lambda} \beta_{-k\lambda}\right) \tag{17}$$

Let us try to diagonalize this. Clearly the only way is to mix, in a way which is by now familiar from superconductivity theory, the operators $\beta^\dagger_k$ with β_{-k} and vice-versa. [Actually, the corresponding transformation was probably first made by Bogolyubov in the theory of He_4 (87).] Let us then set

$$\beta^\dagger_k = u_k B^\dagger_k - v_k B_{-k} \qquad \beta_k = u_k B_k - v_k B^\dagger_k$$

$$\beta^\dagger_{-k} = u_k B^\dagger_{-k} - v_k B_k \qquad \beta_{-k} = u_k B_{-k} - v_k B^\dagger_k$$

To ensure that the new operators have boson commutation rules, we need only check that

$$\left[\beta_k, \beta^\dagger_k\right] = 1 \quad \text{if} \quad \left[B, B^\dagger\right] = 1$$

That is,

$$u_k^2 \left[B_k, B^\dagger_k\right] + v_k^2 \left[B^\dagger_{-k}, B_{-k}\right] = 1$$

or,

$$u_k^2 - v_k^2 = 1$$

This is automatically satisfied if we set

$$u_k = \cosh \frac{\theta_k}{2} \qquad v_k = \sinh \frac{\theta_k}{2}$$

Now let us substitute into the hamiltonian (17). A particular set of terms may be written

$$E_k^o\left(\beta_k^\dagger \beta_k + \beta_{-k}^\dagger \beta_{-k}\right) + E_k'\left(\beta_k^\dagger \beta_{-k}^\dagger + \beta_{-k}\beta_k\right)$$

These now become

$$E_k^o\left\{u_k^2\left(B_k^\dagger B_k + B_{-k}^\dagger B_{-k}\right) + v_k^2\left(B_{-k}B_{-k}^\dagger + B_k B_k^\dagger\right)\right.$$

$$\left. - 2u_k v_k\left[B_k^\dagger B_{-k}^\dagger + B_{-k}B_k\right]\right\}$$

$$+ E_k'\left\{\left(u_k^2 + v_k^2\right)\left(B_k^\dagger B_{-k}^\dagger + B_{-k}B_k\right)\right.$$

$$\left. - u_k v_k\left(B_k^\dagger B_k + B_{-k}B_{-k}^\dagger + B_{-k}^\dagger B_{-k} + B_k B_k^\dagger\right)\right\}$$

Thus we have diagonalized the hamiltonian if we set

$$\frac{2u_k v_k}{u_k^2 + v_k^2} = \tanh\theta_k = \frac{E_k'}{E_k^o} \qquad (18)$$

The remaining terms can be written, when we use

$$B_k B_k^\dagger = B_k^\dagger B_k + 1$$

as

$$\left[E_k^o\left(u_k^2 + v_k^2\right) - 2u_k v_k E_k'\right]\left(n_k + n_{-k}\right)$$

$$+ 2\left(v_k^2 E_k^0 - u_k v_k E_k'\right)$$

The new energy of the excitons, then, is

$$E_k = E_k^o \cosh\theta_k - E_k' \sinh\theta_k$$

But by (18) we may set

$$E_k^o = E_k \cosh \theta_k$$

$$E_k' = E_k \sinh \theta_k$$

checking that

$$E_k = E_k \left(\cosh^2 \theta_k - \sinh^2 \theta_k\right)$$

and then

$$E_k = \frac{E_k^o}{\cosh \theta_k} = E_k^o \left(1 - \tanh^2 \theta_k\right)^{1/2} = \left\{\left(E_k^o\right)^2 - \left(E_k'\right)^2\right\}^{1/2}$$

All of this is remarkably similar to the maneuvers one goes through in diagonalizing the B.C.S. hamiltonian (88), except that for bosons hyperbolic rather than trigonometric angular functions are relevant, and the sign under the square root is correspondingly different. An amusing exercise has been carried out by Suhl (private communication) in diagonalizing this type of hamiltonian in terms of boson-pair operators

$$\beta_k^\dagger \beta_{-k}^\dagger,\ \beta_k^\dagger \beta_k + \beta_{-k}^\dagger \beta_{-k},\ \beta_{-k}\beta_k$$

These boson-pair operators have the properties of spins in hyperbolic rather than real space—i.e., spins with imaginary S, corresponding to the Legendre functions of imaginary order.

We also see that, corresponding to the reduction in frequency of the excitons, there is a decrease in the total energy which may be interpreted very directly as the appropriate lowering of the zero-point energy of the excitons. Using the identity

$$v^2 = \frac{u^2 + v^2 - 1}{2} \qquad \text{from } u^2 - v^2 = 1$$

we obtain

$$\Delta E = \sum_k \left(E_k - E_k^o\right) \tag{19}$$

so that the total diagonalized hamiltonian is

$$\mathcal{H} = \sum_{k>o,\lambda} \left\{ \left(E_0 + X^2 T_{k\lambda}\right)^2 - \left(X^2 T_{k\lambda}\right)^2 \right\}^{1/2} \left(n_{k\lambda} + \frac{1}{2} + n_{k\lambda} + \frac{1}{2}\right)$$

$$- \sum_{k\lambda} \left(E_0 + X^2 T_{k\lambda}\right)$$

where $n_{k\lambda} = B^{\dagger}_{k\lambda} B_{k\lambda}$. Note that $E_k = (E_o^2 + 2E_o X^2 T_k)^{1/2}$. This is clearly the same as (15').

In this representation we have made clear the rather amusing fact, which was pointed out by Hopfield, that the binding energy of molecular crystals may be reinterpreted as the decrease in zero-point energy of excitons due to dipolar interaction, since it is predominantly Van der Waals' attraction.

One rather important point about this zero-point energy problem deserves some discussion. In our original hamiltonian we had, aside from the purely atomic term $(E_o/2)(c_p^{\dagger}c_p - c_s^{\dagger}c_s)$, only terms referring to exciton operators on different atoms, which of course automatically commuted. We chose always to take them in the order $b_j^{\dagger}b_k$, and as a result we came out with a hamiltonian involving $E_k^o \beta_k^+ \beta_k$. The result is the subtractive term above [in (19) the term $-E_k^o$]. But this, of course, while a correct description of our actual physical situation, is not the correctly quantized hamiltonian of a classical dielectric continuum, which should come out in the symmetrized form:

$$\frac{E_k^o}{2}\left(\beta_k^{\dagger}\beta_k + \beta_k\beta_k^{\dagger}\right)$$

Since our assumed system has no obvious physical resemblance to a classical dielectric continuum, this need not worry us much, but it turns out in fact to be all right anyhow. If we wish to use the symmetrized form we must add on $\Sigma_k E_k^o$, which, because of the fact that $\Sigma_k T_k = 0$, depends only on the purely atomic quantity E_o and not on the interactions at all. Of course the reason this identity is true is just that we had no j=k interaction terms, which is why we could choose the order arbitrarily in the first place. In spin-wave and phonon problems this relationship between the classical and quantum starting points will be more to the point.

Now let us begin to move away from the subject of strictly localized Frenkel excitons and to take into account the possibility that the excited electron or hole may occasionally wander away from its partner, in spite of the influence of the Coulomb field of the other. This means that we must include in the hamiltonian the terms

$$\mathcal{H}' = \sum_{jk}\left[b_{jk}^s c_j^{s\dagger} c_k^s + b_{jk}^p c_j^{p\dagger} c_k^p\right] \tag{20}$$

but we must also take into account the energy difference U which decreases the excited electron's energy when it is on the same atom as the hole:

$$\mathcal{H}_{int} = U \sum_j n_j^p n_j^s \tag{21}$$

The best we can do with the terms (20) is to treat them in perturbation theory, starting from the limit in which $U \to \infty$ and finding the first correction of order b^2/U. In the limit of very large U, the appropriate thing to do is to look at the eigenstates of the Coulomb interaction (21). These form themselves into a hierarchy, according to whether 0, 1, 2, etc., pairs of s and p electrons are on the same atom, with energies 0, U, 2U, etc. So far we have treated only the $U = 0$ degenerate subspace of such states, in which the electron and hole occupy the same atom; these have been split up by the dipolar interaction, which is clearly small compared to U. The next stage is to calculate the further splitting of this degenerate subspace by virtual transitions in which the electron or hole hops onto an atom on which there is already another, which in the limit of U very large has energy U. To get the lowest-order effect, clearly we cannot permit a further wandering of the electron but must force it to rejoin its hole on the next hop. We project, then, the perturbation series

$$\mathcal{H}_{\text{eff}} = \mathcal{H}' - \mathcal{H}' \frac{1}{U} \mathcal{H}' + \mathcal{H}' \frac{1}{U} \mathcal{H}' \frac{1}{U} \mathcal{H}' \cdots$$

onto the subspace $|0\rangle$ of "allowable" states in which electron and hole are together:

$$\mathcal{H}^{(2)} = - \langle 0 | \mathcal{H} \frac{1}{U} \mathcal{H}' | 0 \rangle$$

and neglect all higher-order perturbations. We can easily accomplish this by keeping only the terms in which the electron or the hole returns to the atom from which it came, so that the second-order effect of (20) and (21) is

$$\mathcal{H}^{(2)} = - \frac{1}{U} \sum_{jk} \left[b^s_{jk} (c^s_j)^\dagger c^s_k + b^p_{jk} (c^p_j)^\dagger c^p_k \right]$$

$$\times \left[b^s_{kj} (c^s_k)^\dagger c^s_j + b^p_{kj} (c^p_k) \; c^p_j \right]$$

$$= - \frac{1}{U} \sum_{jk} \Big\{ |b^s_{jk}|^2 n^s_j (1 - n^s_k) + |b^p_{jk}|^2 n^p_j (1 - n^p_k)$$

$$+ b^s_{jk} b^p_{kj} (c^s_j)^\dagger c^s_k (c^p_k)^\dagger c^p_j + \text{c.c.} \Big\}$$

The first two terms are a bit unsymmetrical, so we symmetrize between the pairs jk and kj. Also we take into account the fact that within the manifold $|0\rangle$ of states we are considering, $n^p_j = 1 - n^s_j$ and vice versa—either an excited or ground-state electron is on each atom. We then obtain, for these two terms, the result

$$- \frac{1}{U} \sum_{jk} \frac{|b^s_{jk}|^2 + |b^p_{jk}|^2}{2} \left(n^s_j n^p_k + n^p_j n^s_k \right)$$

In terms of the exciton operators,

$$n^s_j n^p_k + n^p_j n^s_k = \frac{1}{2} \left(n^s_j + n^p_j \right) \left(n^s_k + n^p_k \right) - \left(n^p_j - n^s_j \right) \left(n^p_k - n^s_k \right)$$

$$= \frac{1}{2} \left(1 - \sigma^j_z \sigma^k_z \right)$$

or

$$= 2 \left(- n_j n_k + n_j + n_k \right)$$

The latter form shows that, aside from giving a nonlinear interaction between neighboring excitons, this term lowers the exciton energies by

$$\frac{1}{U} \sum_{K} \left\{ |b^s_{jk}|^2 + |b^p_{jk}|^2 \right\}$$

The σ form will be significant in spin-wave problems.

This diagonal term in the energy of the exciton is not of great importance; it reflects the fact that the b term spreads the free-particle levels out into a band, pushing down the mean exciton energy. The other term is a more interesting one; it has precisely the form of the dipolar terms which led to exciton motion, so that we may write these terms in the analogous form to the dipolar ones:

$$\mathcal{H}^{(2)} \simeq - \sum_{jk} T_{jk}' \, b^\dagger_j b_k$$

$$T_{jk}' = \frac{b^s_{jk} b^p_{jk}}{U} \tag{22}$$

The mechanism involved is clearly the virtual hopping of the electron to the neighboring atom, followed by the subsequent hop of the hole in the same direction. This mechanism for transfer is not of much importance for optically active Frenkel excitons but is quite important for forbidden ones, and particularly for spin waves. Clearly the effect of this is to make the exciton spectrum follow slightly the broadening of the single-particle spectrum caused by b_{jk} (Figure 46). As b becomes at all comparable with U this model completely fails. As b increases, we see that

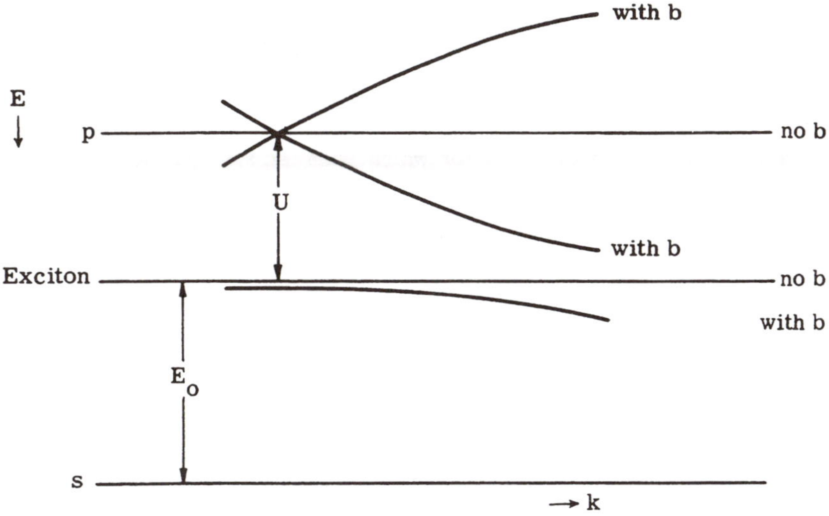

Figure 46

the energy of the exciton states decreases by $\sim Zb^2/U$. On the other hand, the energies of the lowest band states, those which take advantage of the proper phase relationships with their neighbors best, decrease as $\sim Zb$. Thus the *binding energy* of the excitons in this model decreases like

$$BE \cong U - Zb + \frac{Zb^2}{U}$$

so that as soon as $bZ \sim U$, the purely localized excitons will have energies not necessarily below the band edges (Figure 47), and we shall have to make up the

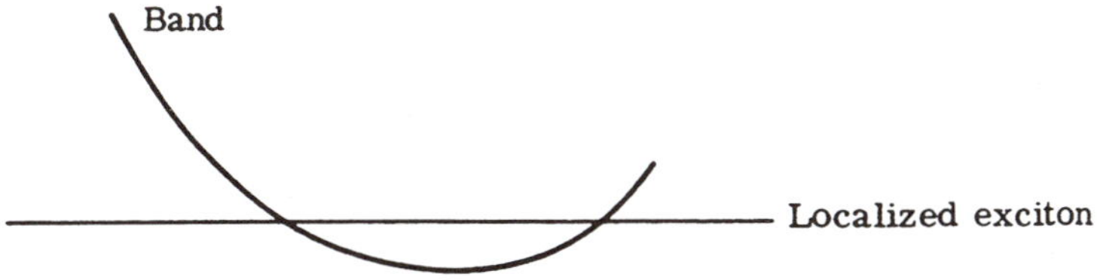

Figure 47

exciton states out of electron states only near the energy minimum, if we are to hope to get an exciton below the continuum of band levels. In addition, in a material in which b is appreciable, the Wannier functions are more extended and dielectric screening will decrease the effectiveness of U. Thus we expect a rather sharp transition, as a function of b/U, to the opposite case of exciton theory, the Wannier limit, in which electron and hole are seldom on the same atom at the same time, but rather have a long-range wave function which can be made up from states near the band minima.

This case can be treated by the "effective one-band hamiltonian" method of band theory — in fact, it was the problem for which Wannier started the modern development of this theory in 1937 (78). We use the approximate hamiltonian

$$\mathcal{H}_{\text{eff}} = E_{\text{el}}(k_e) + E_{\text{hole}}(k_h) + \frac{e^2}{\kappa|(R_e - R_h)|}$$

In general, this leads to rather complicated equations, even though tractable in principle. This is especially so in the case of degenerate bands, where even the center-of-mass motion cannot be separated out easily. Fortunately, the most interesting cases are those in which excitons are observed reasonably easily, and these seem to be simple effective mass ellipsoid situations, where we can write

$$\mathcal{H} = -\frac{\hbar^2}{2} \sum_{\alpha = x,y,z} \frac{1}{m_{\alpha e}} \frac{\partial^2}{\partial x_{\alpha e}^2} + \frac{1}{m_{\alpha h}} \frac{\partial^2}{\partial x_{\alpha h}^2} - \frac{e^2}{\kappa|R_e - R_h|}$$

and where the center-of-mass and relative motions can be separated out, giving us an exciton mass tensor $(m_e + m_h)_\alpha$ for total motion, and a reduced mass tensor

$$\left(\frac{m_e m_h}{m_e + m_h}\right)_\alpha$$

for the relative motion. Even where this is not possible — degeneracy, or ellipsoids with different principal axes — often one particle is much heavier than the other and good approximations can be found.

The exciton wave function in coordinate representation, then, is

$$\Psi = \phi(R_e - R_h)\, e^{iq(R_e + R_h)}$$

if q is the wave number of center-of-mass motion. ϕ is a rather long-range function relative to atomic distances. It is interesting to contrast the approximations made in this treatment with those of our previous Frenkel treatment, and to show graphically how the exciton moves in the two cases. Call the creation operators for Wannier functions in the two bands, electron (conduction) and hole (valence)

$w_e^\dagger(R_j)$, $w_h^\dagger(R_j)$, respectively,

where, for example, w_e is defined as

$$w_e^\dagger(R_j) = \int a_e(r-R_j)\,\Psi^\dagger(r)\,dr$$

e being the empty band. The wave function Ψ may be written

$$\Psi = \sum_{jk} \phi_q(R_j-R_k)\, e^{iq\cdot(\mathbf{R}_j+\mathbf{R}_k)}\, w_e^\dagger(\mathbf{R}_j)\, w_h(\mathbf{R}_k)\Psi_g$$

while the approximate hamiltonian amounts to keeping as interaction energy only

$$\frac{e^2}{\kappa|R_j-R_k|}\, n_e(R_j)n_h(R_k)$$

ignoring all off-diagonal elements. Thus in this case the interaction causes no motion at all: $w_e^\dagger(j)\, w_h(k)\, \Psi_g$ is an eigenfunction of this interaction with energy

$$-\frac{e^2}{\kappa|R_j-R_k|}$$

The motion instead comes entirely from the b_{jk} parts of the hamiltonian. We may describe this by means of a diagram giving time, t, versus position, x, by indicating an interaction taking place at fixed positions of electron and hole, the two moving in x space between interactions (Figure 48). The interactions occur at long range

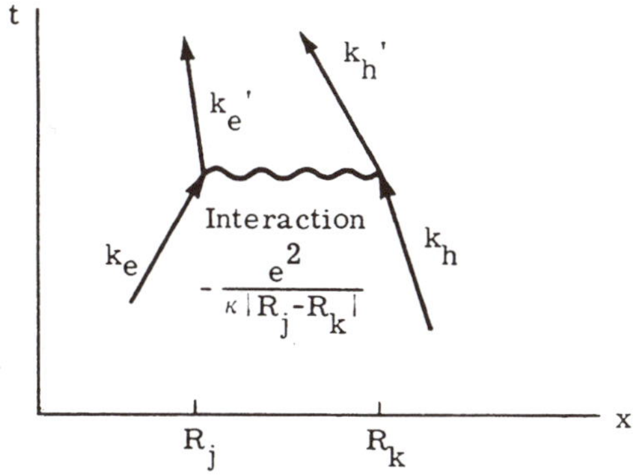

Figure 48

only, and have as their only effect holding the electron and hole together. In the other case, we assumed that b_{jk} was zero and the electron and hole could not move at all, except that w_h and w_e^* could annihilate each other with matrix element X [see Figure 49(a)]. In Figure 49(b) we show the motion described by equation (22)

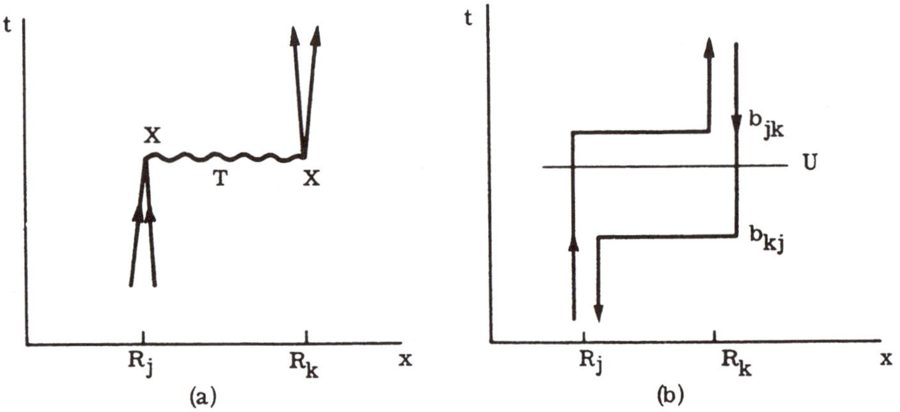

Figure 49

via a virtual jump of electron or hole; this is closely allied to the process of Figure 48. The kind of diagram shown in Figure 49(a) occurs for Wannier excitons also, but only in the rare event that the paths of electron and hole cross (Figure 50). That diagram indicates the fact that the effective dipole moment for interaction with an external field, which is diagrammatically as shown in Figure 51, is reduced by

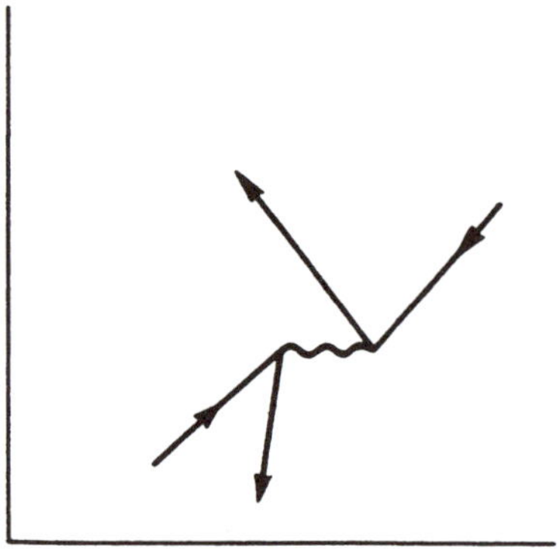

Figure 50

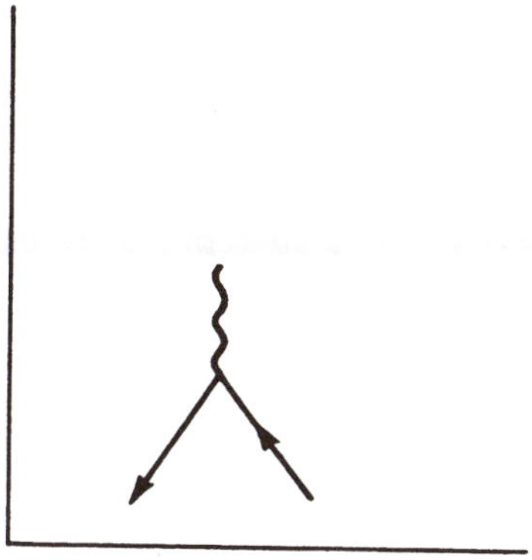

Figure 51

the inverse of the volume of the exciton state, because annihilation only occurs when electron and hole are near each other. The resulting longitudinal-transverse frequency difference is rather small. It cannot be handled in terms of the effective Hamiltonian theory alone.

One way of looking at the zero-point corrections which we made to the ground-state energy and to the frequency of the Frenkel excitons is to observe that they amount to the inclusion of the so-called "backwards diagrams" in the equations for propagation of the excitons (see Figure 52). The "backwards diagrams" have a much larger effect in nuclear many-body theory and in nuclear level structure (89). These diagrams illustrate the process of summing subsets of diagrams in the many-body theory.

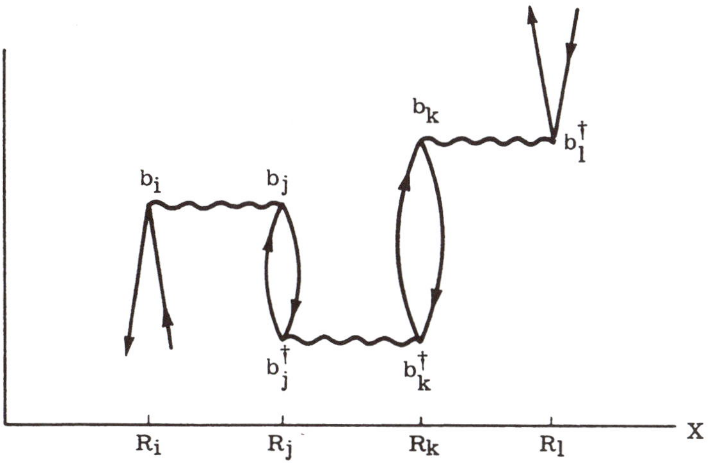

Figure 52

Diagrams like those in Figure 48 are called "ladder diagrams," and solving a Schrödinger equation for two particles is equivalent to "summing the ladders to infinite order"— that is, to allowing the interaction-at-a-distance to operate as often as necessary, essentially continuously. Diagrams like Figure 49 are often called "bubble" or "ring" diagrams, and are related to the dielectric constant of the material. In the simple case of excitons, there is nothing really preventing our summing both sets together except a certain degree of mathematical complexity, and we have given limiting cases in which one type of summing or the other is the more important. In more complicated cases of collective excitations it may only be possible to "sum" the ladder or bubble diagrams alone. In the case of "bubble" diagrams it is important usually also to sum the "backward" ones; this is done by the random-phase approximation (90).

In the case of the Wannier type of exciton there is no particularly obvious or simple way to make the transition to boson operators directly. Instead, we must follow the more ad hoc procedure of writing down the exciton operators and verifying directly that their commutation rules, *in a state in which very few are excited,* are approximately those of bosons. The operator which creates a Wannier exciton may be written

$$O^{\dagger}(q) = \sum_{jk} \phi_q(j-k)\, e^{iq\cdot(R_j+R_k)}\, w_e^{\dagger}(R_j)\, w_h(R_k)$$

We then have

$$\left[O(q),\, O^{\dagger}(q')\right] = \sum_{jklm} \phi_q(j-k)\phi_{q'}(l-m)\, e^{iq\cdot(R_j-R_k) - iq'(R_l+R_m)}$$

$$\times \left[w_h^{\dagger}(R_l)\, w_e(R_m),\; w_e^{\dagger}(R_j)\, w_h(R_k)\right]$$

The commutator will vanish when applied to the ground state Ψ_g unless $R_j = R_m$, $R_k = R_l$ because there are no excitons present *ab initio* in the ground state. That means that $n_h = 1$, $n_e = 0$, i.e., $w_h^{\dagger} w_e = 0$ unless a hole is present on l *and* an excited electron on m. The commutator has the value unity in that case. In the exciton case even physically one seldom reaches a state in which this assumption is invalid, except possibly in certain optical masers. Thus

$$\left[O(q),\, O^{\dagger}(q')\right] = \sum_{j,k} \phi_q(R_j-R_k)\, \phi_{q'}(R_j-R_k)\, e^{i(q-q')\cdot(R_j+R_k)}$$

$$= \delta(q, q')$$

Thus it is quite valid to treat the Wannier excitons as bosons. The most exhaustive treatment of the question in general, of whether and when excitations composed of fermion pairs may be treated as bosons, is in recent work of Blatt and Matsubera which is stiff reading but does seem to settle the fact that they may almost always be so treated.

2. Spin Waves: Heisenberg Hamiltonian and the Magnetic State

Spin waves in insulators are no more than a modification and generalization of the Frenkel exciton. They are an exciton in which the unoccupied or "upper" band is not a distinct orbital band in the Hartree-Fock potential, but the same band as that from which the electron has been excited, but with opposite spin. To understand how this can come about, we must discuss briefly the structure of the "magnetic state" in insulators (91).

Let us start with the simplest case, that of the ferromagnetic insulator; that is, we suppose that a Hartree-Fock solution— not necessarily the lowest one, but self-consistent—has all the states in band n with spin up full, all states with spin down empty.

$$\Psi_{HF} = \prod_k (c^n_{k\uparrow})^\dagger \Psi_{vac} = \prod_j (w^n_{j\uparrow})^\dagger \Psi_{vac}$$

where these two expressions are equivalent (c is the creation operator for Bloch functions, w for Wannier functions) because the band is full.

You will remember that in the case of Frenkel excitons we started out as a zero'th approximation with an atomic excitation energy E_0, which was less than the band excitation energy $E_e - E_h$ by the quantity U which represented the repulsive energy between the two electrons in upper and lower band Wannier functions on the same atom.

Leaving aside the interatomic interactions, E_0 must be zero in the magnetic case, because the excited electron goes into the same orbital as the ground state one. That is, at this stage it costs no energy to turn over a spin *in situ*, but only to turn over the spin and at the same time move the electron to a neighboring atom. If the Wannier function is $a(r-R_j)$, the energy U is

$$U = \int \frac{e^2}{|r_1 - r_2|} dr_1 \, dr_2 \, |a(r_1)|^2 |a(r_2)|^2$$

Clearly the first prerequisite that the magnetic state make any sense at all is that U be large, particularly relative to the "hopping" or kinetic-energy integral

$$b_{jk} = \int a^*(r-R_j) \mathcal{H} \, a(r-R_k) \, dr$$

That U must be large may be seen by considering, again, the ferromagnetic case, taking into account only these two terms in the energy:

$$\mathcal{H}_0 = \sum_{jk\sigma} b_{jk} w_{j\sigma}^{\dagger} w_{k\sigma} + U \sum_j n_{j\uparrow} n_{j\downarrow}$$

We may treat H_0 in Hartree-Fock approximation, assuming that we fill all the states with spin up so that $\langle n_{j\uparrow} \rangle = 1$, $\langle n_{j\downarrow} \rangle = 0$. The energy of a hole state in this potential is of course just

$$E(k) = \sum_j e^{ik \cdot R_j} b_{oj}$$

while that of an electron state is U + E (k) (Figure 53). The situation can obviously

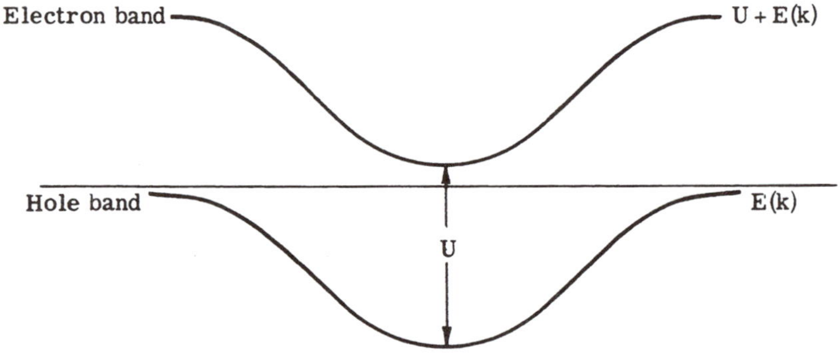

Figure 53

be stable only if all electronic states are above all hole states, i.e.,

$$\left[E(k)\right]_{max} < U + \left[E(k)\right]_{min}$$

This is a necessary condition, but far from a sufficient one, that the electrons not just spontaneously fill up the down-spin states from the up-spin ones and thereby convert the ferromagnetic insulator to a nonmagnetic metal.

Now we can study the stability of the ferromagnetic insulator from a second point of view. That is, we can compute the energies of "excitons" or "spin excitation waves" formed by exciting an electron from an up to a down spin state and allowing the resulting bound complex to travel through the lattice. We shall find that the k = 0 exciton — let us use the term "spin wave" — always has zero energy, but that spin waves of finite wave number may have positive or negative energies. If they are negative, this too represents an instability of the ferromagnetic system, presumably in the direction of a reorientation of the spins in an antiferromagnetic array instead.

This approach to the problem is that of Slater (92) and of Schubin and Vonsovsky (93). Though historically the first real many-body approach, it does not elucidate all the information which can be obtained on the system. More complete and essentially equivalent in rigor is the "vector model" approach, which we shall discuss shortly.

The Coulomb interaction does not contain any direct dipolar matrix elements because the spin-flip transition is doubly forbidden, by spin and parity. Instead, excitation can only be transferred from one atom to the other by the so-called "exchange" matrix elements

$$J_{jk} = \int a^*(r-R_j)\, a(r-R_k) \frac{e^2}{|r-r'|} a^*(r'-R_k)\, a(r'-R_j)\, dr\, dr'$$

which are the Coulomb self-energy of the "overlap charge density"

$$\rho_{jk} = a^*(r-R_j)\, a(r-R_k)$$

Note that because J is a Coulomb self-energy it is necessarily positive.

J multiplies two different groupings of fermion operators:

$$\mathcal{H}_{\text{direct exchange}} = \frac{1}{2} \sum_{jk} J_{jk} \left\{ \left[w^\dagger_{j\uparrow} w^\dagger_{k\uparrow} w_{j\uparrow} w_{k\uparrow} + w^\dagger_{j\downarrow} w^\dagger_{k\downarrow} w_{j\downarrow} w_{k\downarrow} \right] + \left[w^\dagger_{j\uparrow} w^\dagger_{k\downarrow} w_{j\downarrow} w_{k\uparrow} + w^\dagger_{j\downarrow} w^\dagger_{k\uparrow} w_{j\uparrow} w_{k\downarrow} \right] \right\}$$

The first pair of terms may be re-written

$$- (n_{j\uparrow} n_{k\uparrow} + n_{j\downarrow} n_{k\downarrow})$$

$$= -\frac{1}{2} \left\{ (n_{j\uparrow} + n_{j\downarrow})\, (n_{k\uparrow} + n_{k\downarrow}) + (n_{j\uparrow} - n_{j\downarrow})\, (n_{k\uparrow} - n_{k\downarrow}) \right\}$$

$$= -\frac{1}{2} \left(1 + \sigma_{zj} \sigma_{zk} \right)$$

if we assume that there is always one electron per atom:

$$n_{j\uparrow} + n_{j\downarrow} = 1$$

We have made the identification of the fermion pair with a spinor, which in this case is exactly what we usually mean by the spin operator associated with the given atom.

The second pair of terms are more conventional exciton transfer terms; we see that

$$w^{\dagger}_{j\uparrow} w_{j\downarrow} = \sigma_{j+} = \frac{1}{2}(\sigma_{jx} + \sigma_{jy})$$

etc., so that the total exchange energy can be written in terms of spin operators as

$$\mathcal{H}_{\text{direct exchange}} = -\frac{1}{4}\sum_{jk} J_{jk}(1 + \sigma_{zj}\sigma_{jk} + \sigma_{xj}\sigma_{xk} + \sigma_{yj}\sigma_{yk})$$

$$= -\frac{1}{4}\sum_{jk} J_{jk}(1 + \sigma_j \cdot \sigma_k) \tag{23}$$

The advantage of the spin notation is that it allows us at least formally not to confine our thinking to the study of a single spin wave, but to consider the simultaneous reversal of any number of spins in the sample. It depends on the basic assumption, however, that U is very large, as are the excitation energies to all other bands, and thus that we may take $n_{j\uparrow} + n_{j\downarrow} = 1$. Of course this is only exactly true in the unphysical limit $U \to \infty$, but as in the case of Frenkel excitons we may have recourse to perturbation theory projected onto the manifold of states for which it is exactly true. There seems little doubt that such a perturbation theory would converge when b/U is small enough [although Slater has called that into question, for no convincing reason (94)].

On the other hand, so long as we express the results on the interactions of the excitations with each other exactly in terms of the spins σ, there is nothing to prevent us treating situations in which indefinite numbers of spin reversals are present. This is, then, rather a new kind of elementary excitation theory; we allow an indefinite degree of excitation, $n_{exc} \sim N$, but only of excitations of a certain special kind. We may or may not be able to treat the effective hamiltonian (23) of the new theory exactly, but it is a quasi-exact hamiltonian in almost the same sense as that of effective mass theory so long as we refuse to allow real—as opposed to virtual—excitations in which the electrons leave the sites of their corresponding holes permanently.

Within this theory a second and very important mechanism for spin wave motion and interaction is the same perturbation-theoretic term we discussed in the exciton case. A reversed-spin electron may virtually hop to a neighboring atom, giving the spin wave a finite-size wave function. We can simply transcribe the results we obtained in the exciton case. The second-order energy is (taking into account that $b_{j\uparrow k\uparrow} = b_{j\downarrow k\downarrow}$)

$$\mathcal{H}^{(2)} = \frac{1}{U} \sum_{jk} |b_{jk}|^2 \Big\{ n_{j\uparrow} n_{k\downarrow} + n_{j\downarrow} n_{k\uparrow}$$

$$+ w_{j\uparrow} w_{k\uparrow} w_{k\downarrow} w_{j\downarrow} + w_{j\downarrow} w_{k\downarrow} w_{k\uparrow} w_{j\uparrow} \Big\}$$

$$= \frac{1}{2} \sum_{jk} \frac{|b_{jk}|^2}{U} (\sigma_{zj}\sigma_{zk} + \sigma_{xj}\sigma_{xk} + \sigma_{yj}\sigma_{yk} - 1)$$

$$= \frac{1}{2} \sum_{jk} \frac{|b_{jk}|^2}{U} (\sigma_j \cdot \sigma_k - 1)$$

Without any serious study of the spin-wave problem, it is pretty obvious that the ferromagnetic state will be stable if $1/2\, J_{jk} > |b_{jk}|^2/U$ so that the energy is minimized by setting all $\sigma_j \cdot \sigma_k$ as positive as possible, while in the opposite case some antiferromagnetic state will probably be more stable. Let me briefly digress on the physical origin of these, the two most important terms in the exchange interaction in insulators. The term J_{jk} is quite directly the Hartree-Fock exchange term in the interaction energy of electrons in two orthogonal orbitals $a(r-R_j)$ and $a(r-R_k)$ with parallel spin. Because the two electrons have parallel spin, there is an "exchange hole" in the region where their wave functions overlap, which lowers their energy when parallel. This energy gain is responsible for Hund's rule in atoms, the rule that spins of electrons in the same shell will be parallel when possible.

This exchange term is necessarily positive — what, then, of all the arguments which have raged about the signs of exchange integrals? The reason is that the Heitler-London theory of the coupling of two atoms gives as the exchange integral not this alone but an integral involving the full Hamiltonian and arbitrary functions, not necessarily orthogonal, replacing $a(r-R_j)$. Such an integral can have arbitrary sign, and is indeed usually negative. It is this H-L integral which is most often computed. While Herring (95) has shown that the Heitler-London procedure is nearly exactly correct for singly charged atoms at fairly large distances, it seems to me to make the physics clearer to include the primary antiferromagnetic effect as a separate term $\sim -b^2/U$. This term occurs because when the spins of two electrons are antiparallel, the exclusion principle no longer requires them to have precisely orthogonal wave functions. They may gain kinetic energy by spilling over into each others' wave functions.

One can show that when the two wave functions are not automatically orthogonal by symmetry, the b^2/U term almost always predominates; for instance, the *lowest* state of two electrons in any arbitrary ordinary potential is always a singlet, indicating antiferromagnetic exchange, according to a theorem of Wigner (95).

But the working of the exclusion principle can often, as in atoms, cause the wave functions of two electrons to be automatically orthogonal and b_{jk} to vanish. A few isolated cases of insulating ferromagnets, such as CrO_2, obeying this principle, are known by now, but hundreds of other cases exhibit antiferromagnetic exchange.

The usual magnetic materials in which exchange effects are important are oxides or other rather concentrated salts of metals in the iron group or occasionally the other transition groups. A typical case is NiO. In this material the band of interest is the band made up mostly of d states on the Ni ions, although because of covalent bonding of O = p electrons with Ni d states, there is considerable admixture. One finds (91) that reasonable values of b's are ~ 1 ev, of U ~ 10 ev, of J ~ 1/3 of b^2/U.

In this usual case, that $2(|b_{jk}|^2/U) > J_{jk}$, the interaction is referred to as "antiferromagnetic" and the electron spins will tend to make $\sigma_j \cdot \sigma_k$ as negative as possible in the ground state or near T=0. The observed state often approximates reasonably well an array of two or more sublattices of alternating spins (Figure 54). In such a case, of course, we are indeed dealing with a situation which, viewed

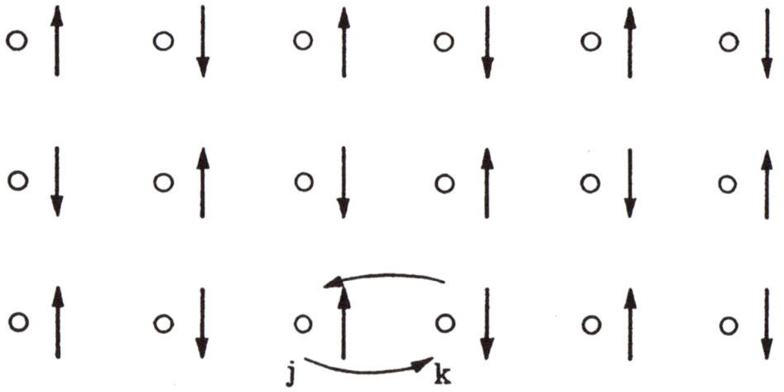

Figure 54

from the ferromagnetic point of view, contains a very large degree of excitation. As we remarked, however, that is quite all right so long as we believe that perturbation theory is valid starting from the quasi-degenerate manifold of states with $n_\uparrow + n_\downarrow = 1$ and allowing no real (as opposed to virtual) transitions into free-electron states.

The criterion for the validity of such a theory must be that b should be considerably smaller than U. In the Frenkel exciton case, you will remember, we had a rapid, but probably not discontinuous, change from the Frenkel to the Wannier type of exciton as b became larger compared to U, but here we can easily see that the transition will be catastrophic.

Just as in the case of Frenkel excitons, the essential question revolves around the binding energy of the excitons. Here, however, we have not got the additional excitation energy E_0 necessary to excite the electron from the s to the p state on the same atom; the only energy preventing the degeneration of the system into a purely metallic state is the binding energy of the electron to its respective hole. By freeing the electron from its hole, we can gain an energy of order $\sim Zb$, but we lose the energy U, and have only a second-order energy $-Zb^2/U$ gained by allowing virtual transitions of the electron away from its hole and vice versa. Thus we may estimate $Zb \sim U$ as a reasonable value of the transition point from antiferromagnet to metal. Additionally, when b is large, again dielectric screening becomes a problem and U is naturally smaller.

Another, also rather approximate, way to do the problem of this transition to the metallic state is to ask whether the alternating array can in fact be a stable Hartree solution of the many-body problem. That is, we can ask what the Hartree field for the down spin at atom k is relative to that at atom j. Nominally the difference is U, the repulsive energy between the up-spin electron at j and the down-spin one. But in a Hartree theory we should take into account the fact that the up-spin electron on j is not actually on j all the time, so that the actual difference in Hartree fields may be considerably smaller.

The electron of up spin on site j hops to each of its neighbors k with probability $|b_{jk}/U|^2$, while equally site k may with probability $|b_{jk}/U|^2$ be occupied by an up-spin electron. Thus the actual difference in energy between a down-spin state on j as opposed to k is

$$\Delta E = U(\langle n_{j\uparrow}\rangle - \langle n_{k\uparrow}\rangle)$$

$$= U\left(1 - 2Z\left|\frac{b_{jk}}{U}\right|^2\right)$$

But actually the energy denominator which determines the amount of admixture should be not U itself but ΔE, so we get an equation for self-consistency

$$\Delta E = U\left(1 - \frac{2}{(\Delta E)^2}\sum_k |b_{jk}|^2\right)$$

or

$$(\Delta E)^3 - U(\Delta E)^2 = -2UZ|b_{jk}|^2$$

This equation may be graphed as shown in Figure 55. Thus the maximum possible value of b is given by

$$\left(\frac{b}{U}\right)_{max} \cong \left(\frac{2}{27Z}\right)^{1/2} \sim \frac{1}{10} \text{ or so}$$

At any b above this critical value, the magnetic state can no longer be maintained. It will be unstable against a redistribution of the electrons so that the occupancy of both spin states is equal, and the state is metallic. The lower iron-series oxides,

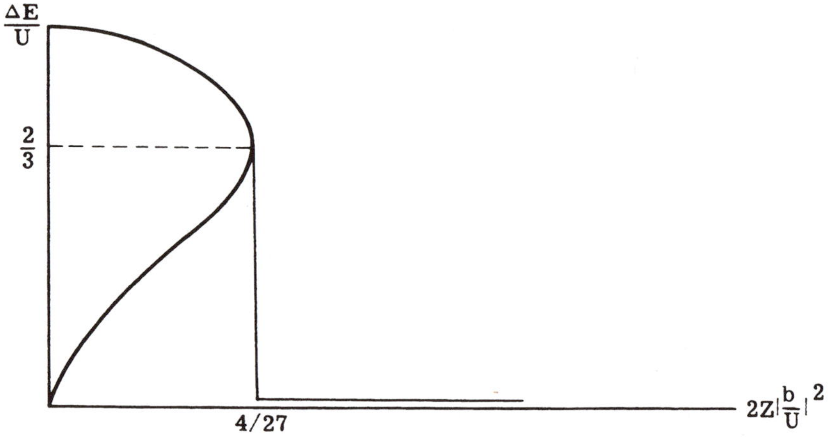

Figure 55

and those of the later transition series — V, Ru, etc., oxides — are metallic, although the mobility of the electrons is quite low, 10^{-3} or so of that in usual metals. Some exhibit a thermal transition which, it has been suggested, may be of the nature of a metallic-to-magnetic transition (9).

3. Ferromagnetic Spin Waves

Now let us go on to the elementary excitations, first in the ferromagnetic case (96). The effective hamiltonian is

$$\mathcal{H} = -\frac{1}{4} \sum_{jk} J_{jk} \, \sigma_j \cdot \sigma_k$$

(where we assume $J_{jk} > 0$).

One of the most important properties of spin vectors, which they have in common with harmonic oscillators, is that their classical equations of motion are identical with the quantum ones. Suppose σ enters into the hamiltonian multiplied by a field H: $\mathcal{H} = -1/2 \; \gamma\hbar \, \mathbf{H} \cdot \boldsymbol{\sigma}$. Then, when we calculate the equation of motion of the operator σ in Heisenberg representation:

$$i\hbar \frac{d\sigma}{dt} = \left[\sigma, \mathcal{H}\right]$$

we obtain

$$i\hbar \frac{d\sigma}{dt} = -\frac{\gamma\hbar}{2}\left[\sigma,\sigma\right]\cdot \mathbf{H}$$

{ This notation means that $[\sigma,\sigma]$ is the dyadic made up of the components $[\sigma_i,\sigma_j]$.}
The commutators of different components are summarized in the relationship

$$\sigma \times \sigma = 2i\sigma$$

or

$$\sigma_i\sigma_j - \sigma_j\sigma_i = 2i\epsilon_{ijk}\sigma_k$$

so that

$$\left[[\sigma,\sigma]\ \mathbf{H}\right]_k = 2i\sum_j H_j\,\epsilon_{kjl}\sigma_l = 2i\,\mathbf{H}\times\sigma$$

Thus

$$\frac{d\sigma}{dt} = \gamma\,(\sigma\times\mathbf{H})$$

which is indeed the classical equation of motion

$$\frac{d\mathbf{S}}{dt} = \gamma\,(\mathbf{S}\times\mathbf{H})$$

of a spin $\mathbf{S} = \sigma/2$ in a magnetic field H and with gyromagnetic ratio γ. H may be a c-number or any operator which commutes with σ, the spin under consideration.

The field **H** which spin σ_j sees is $(1/2\gamma\hbar)\ \Sigma_k J_{jk}\sigma_k$. It is convenient to introduce a preferred direction $\hat{z}$ by also having a small external field, leading to a term in the hamiltonian $\hbar\gamma H\sigma_z$, so that we have the equation of motion

$$\frac{d\sigma_z}{dt} = \frac{1}{2\hbar}\sum_k J_{jk}\sigma_j\times\sigma_k + \gamma H(\sigma_j\times\hat{\mathbf{z}}) \tag{24}$$

These equations can be used to obtain equations of motion of the spin waves which are entirely equivalent to those derived by using boson theory, and many of the properties can be calculated directly using spins and the equations of motion without ever going through the pseudo-boson approximation. Unfortunately, the thermal and energetic properties are not as easily done this way, although modern Green's function theories have made considerable progress in this direction. It is nevertheless very important to remember that in spin theory, as in harmonic oscillator theory, most classical results are not approximations at all, and that one's classical intuition can be of the greatest value.

Let us write out the equation of motion (24) as

$$\frac{d\sigma_{jx}}{dt} = \frac{1}{2\hbar}\sum_k J_{jk}(\sigma_{jy}\sigma_{kz} - \sigma_{ky}\sigma_{jz}) + \gamma\sigma_{jy}H$$

$$\frac{d\sigma_{jy}}{dt} = \frac{1}{2\hbar}\sum_k J_{jk}(\sigma_{jx}\sigma_{kz} - \sigma_{kx}\sigma_{jz}) - \gamma\sigma_{jx}H$$

Adding -i times the second to the first, we get

$$\frac{d\sigma_j^-}{dt} = \frac{1}{2}\frac{d}{dt}(\sigma_{jx} - i\sigma_{jy})$$

$$= \frac{i}{2\hbar}\sum_k J_{jk}(\sigma_j^-\sigma_{kz} - \sigma_k^-\sigma_{jz}) + i\gamma\sigma_j^-H$$

The fact that only σ^-'s enter indicates that spin waves are circularly polarized. Since we are interested in calculating all the properties, thermal as well as dynamic, let us make the boson transformation

$$\sigma_j^- = b_j^\dagger(1-n_j) \qquad \sigma_{jz} = 1 - 2n_j$$

We have reversed the sign because the ground state corresponds to $\sigma_z = +1$, and a spin wave involves turning a spin *down*. The linear terms in the b's are

$$\frac{db_j^\dagger}{dt} = i\gamma H b_j^\dagger + \frac{i}{2\hbar}\sum_k J_{jk}\left(b_j^\dagger - b_k^\dagger\right)$$

When we transform to spin-wave variables β_k we get

$$\frac{d\beta_k^\dagger}{dt} = i\omega_k\beta_k^\dagger$$

$$\omega_k = \gamma H + \frac{1}{2\hbar}\sum_j J_{oj}(1-\cos k\cdot j) \tag{25}$$

In the absence of a magnetic field, the spin-wave frequencies increase as k^2 from the value 0 at k=0. The fact that, aside from the external field, the frequency must be zero at k=0 is the consequence of an identity which comes from the general equation of motion:

$$\frac{d}{dt}\sum_j \sigma_j^- = \frac{i}{2\hbar}\sum_{jk} J_{jk}(\sigma_j^-\sigma_{kz} - \sigma_k^-\sigma_{jz}) \equiv 0$$

and in fact $S_0 = \Sigma_j \sigma_j$, the total spin vector itself, is a constant of the motion. S^- is $\sqrt{N}$ times the k=0 spin-wave operator.

This is the first instance we have come upon of one of the most important general rules about collective excitations, which is related to the presence of symmetry in the hamiltonian. The exchange hamiltonian (23), as well as the original hamiltonian from which it is derived, are invariant under the operation of rotation in spin space. This is because the Coulomb interactions, as well as the kinetic energy, do not involve the spins of the electrons in any way; the spins only enter the problem through the exclusion principle or the more complete statement of fermion statistics, the commutation relations. But the commutation relations themselves are invariant to spin rotations (of all spins simultaneously, of course):

$$\left[\Psi_\sigma(r),\ \Psi_\sigma^\dagger(r')\right] = \delta_{\sigma\sigma'}\,\delta(r-r')$$

is also true after the transformation corresponding to a rotation in spin space

$$\Psi_+ \rightarrow \cos\frac{\theta}{2}\, e^{-i(\phi/2)}\Psi_+' + \sin\frac{\theta}{2}\, e^{i(\phi/2)}\Psi_-'$$

Ψ_+' is then the + component of the wave function when we quantize along θ, ϕ instead of along z (Figure 56).

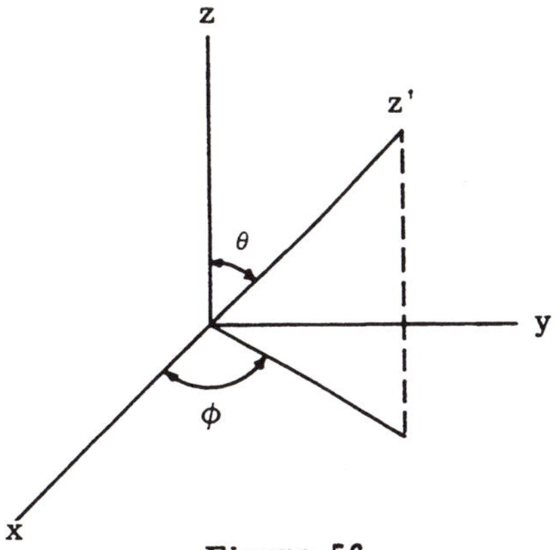

Figure 56

To every invariance of the hamiltonian corresponds a constant of the motion. Corresponding to time-translation invariance, the constant is E; to space-translation invariance, P; and to angular rotation invariance, the total angular momentum L. These are the so-called "generators" of the symmetry transformations; for instance, $L_z = i(\partial/\partial\phi)$ and when $\partial\mathcal{H}/\partial\phi = 0$

$$\frac{\partial}{\partial\phi}\mathcal{H} - \mathcal{H}\frac{\partial}{\partial\phi} = 0 \qquad \text{so that} \qquad [\mathcal{H}, L_z] = 0$$

Thus rotation invariance in spin space implies $[\mathcal{H}, \mathbf{S}] = 0$; $\mathbf{S} = \text{const}$. Actually, this is not an exact symmetry because the magnetic part of the electromagnetic forces does depend on σ; it is only valid in the limit $v/c \to 0$, so that, e.g., spin-orbit coupling violates it. As a result, real ferromagnets have what is called "crystal anisotropy energy," which depends on the direction of **S** in space, but which is almost always minor in magnitude compared to the exchange, which comes from Coulomb energies, and which basically determines the magnetic properties.

This approximate symmetry ensures that the frequency of the k=0 spin wave shall be near zero. What is the excited state which comes from excitation of the k=0 spin wave? The original ferromagnetic state contained precisely N electrons, each of them in the spin-up state, so that the total spin was $S = N/2$ ($\sigma_{tot} = N$), pointed precisely in the z direction. Because of the rotational invariance of the hamiltonian, the z direction is of course entirely arbitrary, and we could have quantized along the x or y direction, or any other, if we had liked. Corresponding to this, there is a degeneracy in the ground state; sticking to the z direction of quantization, we could have chosen

$$S_z = -\frac{N}{2}, -\frac{N}{2} + 1, \cdots, 0, 1, \cdots \frac{N}{2} - 1, \frac{N}{2}$$

and each of these exact eigenstates has just the same energy as the original state. By taking the appropriate linear combinations of these, we could make up a state with $S_{z'} = N/2$, with z' pointing in any direction θ, ϕ. What linear combination to take can be calculated from the appropriate representation coefficient $D^{N/2}_{N/2,\,N/2}(\theta, \phi)$ or by the trick of considering the state N to be made up of [$\alpha(j)$ is the spin-up component of wave function corresponding to σ_j]

$$\alpha(1)\alpha(2)\cdots\alpha(N)$$

and using the transformation

$$\alpha \to \alpha' \cos\frac{\theta}{2}\, e^{i(\phi/2)} + \beta' \sin\frac{\theta}{2}\, e^{-i(\phi/2)}$$

It is obvious that the states obtained by successive excitations of the k=0 spin wave from the state $S_z = N/2$, $S = N/2$, are the states $S = N/2$ but $S_z = (N/2) - 1$, $(N/2 - 2$, etc. This is because

$$\beta^{\dagger}_{k=0} = \frac{1}{\sqrt{N}} \sum_j \sigma_j^- = \frac{1}{\sqrt{N}} S^-$$

and S^- is the "lowering operator" which decreases S_z by unity without changing $|S|$. Thus we see that the rotational invariance of the ground state and the zero frequency of the k=0 mode are equivalent.

The eigenstates reached by the k=0 mode excitation are the same even in the presence of a magnetic field, since S_z and $|S|$ are still good quantum numbers in the presence of a perturbation $\hbar\gamma HS_z$. Then the frequency of the k=0 spin wave is not zero but exactly γH. Note that these statements about the frequency of the uniform mode are true, in the absence of spin-orbit forces, for all states of the hamiltonian; this is obvious because S^- leads to a mere rotation of any state. This is the famous theorem that "exchange alone does not broaden the magnetic resonance frequency." Another way to put it is that the k=0 spin wave does not interact in any way with the $k \neq 0$ waves.

Now, what about $k \neq 0$ spin waves? It is an obvious hypothesis, and does, in fact, turn out to be true, that when only exchange forces are considered, the spin waves will have frequencies joining continuously on to that of the k=0 mode—that is, $\omega \to 0$ as $k \to 0$.

This is not obvious, in view of the complications of the $k \to 0$ limit in other problems, and is a result of the short-range nature of exchange forces. For instance, when we take the long-range magnetic dipolar forces into account, the k=0 mode's frequency depends on the shape of the sample because the demagnetizing effects caused by surface poles depend on sample shape. Also the frequencies of longitudinal and transverse spin waves differ, so that there is no question of continuity as $k \to 0$. But, in contrast, exchange forces are basically short range, because aside from magnetic dipolar and spin-orbit effects, a redistribution of spin moment does not affect the charge distribution in any way perceptible at long range.

This, then, is the important connection which I should like to emphasize, because it plays a vital role in the general theory of collective excitations and of phase transitions. We start from a hamiltonian

$$\frac{1}{4} \sum_{jk} J_{jk}\, \sigma_j \cdot \sigma_k$$

which has perfect rotational symmetry in spin space. In spite of this rotational symmetry, the lowest eigenstate state which we postulate—and which, under special conditions, namely all $J_{jk} < 0$, *is* really the lowest eigenstate—does not have perfect rotational symmetry. Only singlet states, $S_{tot} = 0$, have that. Therefore the lowest eigenstate must be degenerate with a large number of other states, pointing in different directions or, if we like, with different S_z along the same direction.

As a result, we conclude that there is at least one form of elementary excitation, that corresponding to k=0, which gives a rotation of the entire system in spin space, and which must have frequency $\omega = 0$. Under one additional physical condition, the absence of long-range forces, this implies that there is a whole branch of the spectrum of elementary excitations which has $\omega \to 0$ as $k \to 0$.

Spin waves are by now an experimental phenomenon as thoroughly tangible as phonons. They have been studied by neutron inelastic scattering, by direct observation of their magnetic resonances in thin films, by coupling with the phonon spectrum, and a number of other more indirect ways. Their best-known property, however, is their effect on the temperature variation of magnetization and of specific heat. A branch of boson excitations of frequency $\hbar\omega = Ak^2$ has a number density

$$N = 4\pi \int_0^{k_{max}} \frac{k^2 dk}{e^{\beta A k^2} - 1} \qquad \beta = \frac{1}{k_B T}$$

and, when $Ak_{max}^2 \gg k_B T$, as at low temperatures, this is clearly

$$N \simeq 4\pi \int_0^{\infty} \frac{k^2 dk}{e^{\beta A k^2} - 1} = \frac{4\pi}{(\beta A)^{3/2}} \int_0^{\infty} \frac{\chi^2 dx}{e^{\chi} - 1}$$

$$= \frac{4\pi (k_B T)^{3/2}}{A^{3/2}} \times \text{const.}$$

Now we observe that the number of reversed spins per spin wave is precisely 1, so that the change in S_z is just $\propto T^{3/2}$, the famous temperature variation of the Bloch theory. Dyson (81) has computed correction terms to this variation; there are terms in $T^{5/2}$ and $T^{7/2}$ coming from deviations from $\hbar\omega = Ak^2$, but the first spin-wave interaction term is a completely negligible one of order T^4. We would at first expect $A(T) \propto$ const. - N, so that we expect a T^3 term, but the interaction of spin waves of long wavelength is very small, essentially because of continuity with the $k=0$ wave, which has no interaction with other spin waves; one finds

$$\omega \propto k^2 \left[\text{const.} - \int [k'^2 N(k')]\, k'^2 dk' \propto k^2 [\text{const.} - T^{5/2}]\right.$$

where $N(k)$ is the average number of spin waves of wave vector k present. Correspondingly, the correction term is $T^{3/2}\, T^{5/2} = T^4$. The spin-wave specific heat is also $\propto T^{3/2}$: the energy is

$$U = 4\pi A \int \frac{k^2}{e^{\beta A k^2} - 1} k^2 dk \propto T^{5/2} \qquad C = \frac{\partial U}{\partial T} \propto T^{3/2}$$

An important point is illustrated in the behavior of these integrals near $k=0$. At any finite temperature, near $k=0$, the spin-wave density $N(k)$ is the classical value

$$N(k) = \frac{k_B T}{\hbar\omega(k)} = \frac{k_B T}{Ak^2}$$

Notice that if the system were only two-dimensional, so that the density of states were $\propto k\,dk$, the integral for the magnetization deviation would diverge logarithmically at $k \to 0$. Any small external field or spin-orbit coupling can, however, make the frequency of the $k=0$ spin-wave finite and the integral converge, essentially as

$$T \ln\left(\frac{k_B T}{\text{cutoff energy}}\right)$$

This two-dimensional divergence is also a very general property of ordered systems with short-range forces. For instance, it is believed a two-dimensional crystal would not be thermally stable in the absence of an ordered support of some sort. However, there apparently exists no actual proof of the absence of a transition in two dimensions (98).

Now I should like to make the connection between these results and those of two different kinds of classical theory. The first is the corresponding problem for purely classical spin vectors. Suppose that instead of the Pauli spin operators σ_j we had classical spin vectors $\mathbf{S}_j$:

$$\mathcal{H} = -\sum_{jk} J_{jk}\, \mathbf{S}_j \cdot \mathbf{S}_k$$

We define $\mathbf{S}_j = J_j/\hbar$, where J_j is the true classical angular momentum. Thus the classical exchange parameter is $J_{jk}/\hbar^2$. We point this out because our artificial use of S rather than J may introduce factors of $\hbar$ into equations which are actually classical.

The lowest state of this hamiltonian would be that in which the classical spin vectors are all perfectly aligned in the same direction. We can derive an approximate hamiltonian in terms of the classical precessional modes of such an array by expanding this energy about the equilibrium position:

$$\mathbf{S}_j \cdot \mathbf{S}_k = S_{zj}S_{zk} + S_{xj}S_{xk} + S_{yj}S_{yk}$$

$$S_{zj}^2 = S^2 - S_{xj}^2 - S_{yj}^2$$

so that

$$S_{zj} \simeq S - \frac{1}{2S}(S_{xj}{}^2 + S_{yj}{}^2)$$

because near equilibrium, the second term is small. Thus

$$-\mathcal{H} = -NZJS^2 + \sum_{jk} J \left[(S_{xj} - S_{xk})^2 + (S_{yj} - S_{yk})^2 \right]$$

$$= -NZJS^2 + \sum_{\substack{jk \\ \text{neighbors}}} \frac{J}{2} \left[(S_j^+ - S_k^+)(S_j^- - S_k^-) + (S_j^- - S_k^-)(S_j^+ - S_k^+) \right]$$

Here we have assumed that J_{jk} vanishes except between an atom and its Z neighbors, an inessential simplification. S^- and S^+ are defined as for quantum mechanical spins. We may proceed further by observing that, classically as well as quantum mechanically, small oscillations $\hbar^{1/2}(\delta S_x/\sqrt{S})$ and $\hbar^{1/2}(\delta S_y/\sqrt{S})$ of the spin obey the Poisson bracket relationships for canonically conjugate variables. More simply, we remark that the classical and quantum equations of motion of spins are identical, so that the classical modes of the system are also precessional waves:

$$S_j^-(t) = e^{i\omega_\lambda t} e^{i\lambda \cdot j}$$

That is, we may derive a set of normal modes

$$S_\lambda^- = \frac{1}{\sqrt{NS}} \sum_j e^{i\lambda \cdot j} S_j^-$$

$$S_\lambda^- \propto e^{i\omega_\lambda t}$$

ω_λ is given by

$$\omega_\lambda = \frac{S}{\hbar} \sum_j J_{oj} (1 - \cos \lambda \cdot j)$$

which agrees with (25) if in (25) we insert $S = 1/2$, and also is a purely classical formula if we substitute $J = \hbar S$ and realize that J_{oj} is $\hbar^2$ times the classical exchange parameter. The hamiltonian reduces to

$$\mathcal{H} = -NZJS^2 + \sum_\lambda \frac{\omega_\lambda}{2} (S_\lambda^+ S_\lambda^- + S_\lambda^- S_\lambda^+) \tag{26}$$

Of course, the order is irrelevant classically, but when we quantize the system we must, as always, choose this symmetrized, Hermitian combination.

The situation with regard to the zero-point motion and energy is at first confusing. It appears as if there is a discrepancy between the quantized classical Hamiltonian and the corresponding quantum results, because quantum mechanically the energy of the ground state is exactly $-NZJS^2$, and there is no room here for the correction by the commutator term which must come from reordering the $S^+S^- + S^-S^+$ term and correcting:

$$\frac{1}{2}\left\{(S_\lambda^x + iS_\lambda^j)(S_\lambda^x - iS_\lambda^y) - (S_\lambda^x - iS_\lambda^y)(S_\lambda^x + iS_\lambda^y)\right\}$$

$$= -i(S_\lambda^x S_\lambda^y - S_\lambda^y S_\lambda^x) = i\frac{1}{NS}\sum_j iS_j^z$$

$$= 1$$

The situation is saved, as was observed first by Klein and Smith (99), by the fact that the classical spin corresponding to a quantum spin vector S is $S' = \sqrt{S(S+1)}$, i.e., $S(S+1) = S_{xj}^2 + S_{yj}^2 + S_{zj}^2$. Thus, $S_{zj}^2 = S^2 - (S_{xj}^2 + S_{yj}^2 - S)$, and again applying the approximate commutator

$$S_{xj}S_{yj} - S_{yj}S_{xj} = iS_{zj} \simeq iS$$

this is

$$S_{zj}^{\ 2} = S^2 - \left[\frac{1}{2}(S_j^+S_j^- + S_j^-S_j^+) - S\right]$$

$$\simeq S^2 - S^-S^+$$

Thus (26) is the correct correspondence-principle hamiltonian only if we change S^2 to $S(S+1)$, and if we do that it agrees exactly with the quantum result

$$\mathcal{H} = -NZJS^2 + \sum_\lambda \omega_\lambda S_\lambda^- S_\lambda^+$$

The result is a perfect correspondence between the classical and quantum theoretical results. Again, as in the exciton case when we neglected the "backward diagrams," the zero-point-motion effect is a rather trivial mathematical exercise relevant only to the internal dynamics of the individual atoms; the coupling terms have a vanishing net zero-point average because the zero-point effect is summed over all spin waves λ and $\Sigma_\lambda \cos\lambda = 0$ ($\cos\lambda$ is, in this case, the Fourier transform of the coupling term).

A second very generally useful semiclassical approach to magnetic problems was suggested by Landau, Lifshitz, Kittel, and Herring (100). This approach is very close to the Debye theory of phonons. The idea was to look at the magnetization from a rather coarse-grained point of view as a continuous magnetization field M(r) obeying the commutation relations of a continuous quantum field,

$$M(r) \times M(r') = iM(r)\ \delta\ (r-r')\ (\times \text{ consts.})$$

The form of the hamiltonian, depending on $(S_j - S_k)^2$, suggests that we use as a semiempirical hamiltonian density

$$\mathcal{H} = A \int dr \left(\nabla M(r)\right)^2 \tag{27}$$

This can be shown to be correct for the special case of the insulating ferromagnets, and it is pretty obviously also the macroscopic consequence of any short-range, isotropic force—just as elastic theory necessarily gives us a hamiltonian quadratic in the deformation (∇u) (u the displacement). From this were derived general magnetization equations of motion valid, one could conjecture, in all ferromagnets:

$$\frac{dM}{dt} = A(M \times \nabla^2 M) + \gamma(M \times H_{ext}) + (\text{dissipative terms})$$

This is the Landau-Lifshitz equation, which plays a basic role in all spin-wave and ferromagnetic resonance theory.

One interesting property of hamiltonians like (27)—and the corresponding hamiltonian of elastic theory,

$$\mathcal{H} = \int \nabla u : C : \nabla u \qquad (\text{C the stiffness elastic tensor})$$

— is the general argument that two-dimensional systems with short-range forces cannot have stable long-range order. We believe readily that the long-wavelength motions of a system, at least when their frequencies are low, behave classically at any finite temperature. Then equipartition allows us $k_B T$ of energy per degree of freedom k, and the energy is

$$\propto k^2 \left(\nabla M(k)\right)^2 \qquad \text{so} \qquad \langle \left(\nabla M(k)\right)^2 \rangle = \frac{k_B T}{k^2}$$

Thus the fluctuation integral

$$\langle \Delta M^2 \rangle = \int k^{D-1}\, dk\, \frac{k_B T}{k^2} \qquad (\text{D the dimensionality})$$

diverges for D = 2, in both magnetic and elastic cases. Another way of putting it is that one may, in two dimensions, create a domain of reversed magnetization of size l anywhere in the sample, surrounded by a "Bloch wall" of thickness $\sim l$, with energy (number of atoms in wall) $\times$ $(\nabla M)^2 \sim l^2 (1/l^2)M^2$ = finite. Thus we have a finite thermal excitation probability for such domains in all areas of the sample, on any scale and in essentially an infinite number of possible configurations. Thus long-range order is destroyed. In general, the Landau-Lifshitz equation allows us to deal with motions and configurations more complex and general than those of spin waves, so long as they are slowly varying.

4. Antiferromagnetic Spin Waves and "Broken Symmetry"

Finally, I shall give a brief introduction to the subject of antiferromagnetic spin waves (101). These are important because they are probably the simplest and best-understood case of the various kinds of relationships among zero-point motion, symmetry, and elementary excitation spectra which we can lump together under the name of the "theory of broken symmetry" and which play a vital role in the theory of almost all forms of quantum condensation—solids and phonons, spin waves, superconductivity, liquid helium, and the plasma.

In general, a condensation phenomenon is characterized by the appearance of a new parameter in the state of what we call the "condensed" system which was not present in the original system. Above its Curie point a ferromagnet has no magnetization in the absence of an external field; below it does. In an antiferromagnet, the new parameter is the antiferromagnetic order. We may use, for instance, the difference in magnetizations of the two sublattices as our parameter—it is often called the "order parameter." Order-disorder in alloys has an obvious new parameter, as does ferroelectricity. Equally obvious is the long-range order of a solid, which is not present in the liquid. Liquid helium II has a parameter which now is generally understood to be the number of atoms in the zero momentum state. The liquid-gas system is a little more evasive, but by using the so-called "lattice gas" method we can make an analogy to a classical order-disorder system. Still more evasive, but exactly similar, is the order parameter of the superconductor, which may be defined as a certain off-diagonal "Green's function" related to the energy gap.

All of these order parameters have one feature in common: When the order parameter is present, the system has a lower symmetry than in the uncondensed phase. Again this is quite obvious in most cases, although for the two superfluids this lower symmetry is rather a peculiar and controversial one, namely a lack of gauge invariance of the usual sort (perhaps more generally accepted is the idea that zero momentum plays a special role, i.e., lack of Newtonian invariance). In most other cases the change in symmetry is of a kind which is less inflammatory to the sensitivities of the average theoretical physicist, because he is more familiar with the existence of perturbations which destroy it. Thus the ideal ferro- and antiferromagnet have discarded the rotational symmetry of the exchange forces in spin space, which is destroyed anyhow by magnetic perturbations. The fact that the ferromagnetic and antiferromagnetic states do not have time-reversal symmetry (a very practical point for ferromagnets, since it permits the existence of ferrite gyrators) should disturb the theorist almost as much as gauge in the superconductor, because time reversal is usually considered to be an exact symmetry. The solid has discarded full rotational and translational symmetry, which again does not disturb our intuition, but only because we live every day with solid clamps and boxes which have the same peculiar lack of symmetry. Nonetheless, in all these cases we use for our hamiltonian, ordinarily, an expression which *has* the appropriate symmetry, and then we derive—or assume we can derive—from it a state of the system which is determined by that hamiltonian but which *does not* have the symmetry. This, then, is a general feature, as far as I can see, of all condensation phenomena.

Ferromagnetism is unique, or nearly so, among all these condensed phases, in that the order parameter in this case is a *constant of the motion,* at least approximately. This fact allows us a great degree of simplification. In the idealized Heisenberg model of the pure ferromagnet, in the first place, we can write down an exact ground state of the system; in more complicated models, such as the Heisenberg ferrimagnet, we can at least characterize the ground state by a good quantum number—the total spin, and/or the total z component thereof. The existence of such a good quantum number implies an exact statement about the symmetry of the ground state. Then we can characterize all excited states by saying that there is at least one reversed spin. The situation is like that in the N+1 body problem, in that the states with elementary excitations can really be distinguished clearly from the ground state. Also the amount of magnetization reversal, like the change in particle number, is a fixed constant per elementary excitation. As we pointed out already, the existence of the symmetry in the hamiltonian which is "broken" in the ferromagnetic state implies, of course, that there are actually a whole collection of ground states, in this case N+1, which are transformed into each other by rotation of the whole spin system. Because the spin is a constant of the motion, these different states are all exact eigenstates of the hamiltonian, and as we saw this implied that $\omega(k=0)$ is identically zero.

In nearly all the other interesting cases of condensation, the order parameter is not a constant of the motion. Let us, for instance, take the simplest case, antiferromagnetism. The hamiltonian of an ideal antiferromagnet of a particularly simple kind is

$$\mathcal{H} = J \sum_{jl} \mathbf{S}_j \cdot \mathbf{S}_l$$

where we may simplify things inessentially by supposing that the S's are on a simple cubic lattice and that J only acts between nearest neighbors, so that S_j may be taken always to be on the face-centered cubic sublattice A, and S_l on B (Figure 57).

Figure 57

The equation of motion of the spin of the A sublattice is

$$\left[\mathcal{H}, \mathbf{S}_A\right] = J \sum_j \mathbf{S}_j \times \sum_{\text{neighbors to } j} \mathbf{S}_l$$

$$= J \sum_{\substack{\text{nearest-neighbor} \\ \text{pairs}}} \mathbf{S}_j \times \mathbf{S}_l = -\left[\mathcal{H}, \mathbf{S}_B\right]$$

where $\mathbf{S}_B$ is the total spin of the other sublattice. Thus, while the total spin $\mathbf{S}_A + \mathbf{S}_B$ is, by our general theorem, a constant of the motion, none of the components of the difference in spin, $\mathbf{S}_A - \mathbf{S}_B$, is:

$$\left[\mathcal{H}, \mathbf{S}_A - \mathbf{S}_B\right] = 2J \sum_{\text{pairs}} \mathbf{S}_j \times \mathbf{S}_l$$

The quantity on the right is emphatically not zero in any possible state of the system near the antiferromagnetic state; in the ferromagnetic state it can be. For instance, the (-) component is

$$\left[\mathcal{H}, S_A^- - S_B^-\right] = J \sum_{j,l} (S_j^- S_{lz} - S_l^- S_{jz})$$

In the state (by no means an eigenstate) in which $S_{jz} = S$, $S_{lz} = -S$, this is

$$\frac{JS}{\hbar} \sum_{jl} (S_j^- + S_l^-)$$

the matrix elements of which are not zero in this state, although, of course, there is no *diagonal* element. In general, without going into any long demonstration I will assure you that $S_A - S_B$ cannot be a constant of the motion in the antiferromagnetic case.

On the other hand, what does unquestionably happen is that the rate of precession of this "order parameter" variable can be extremely slow—negligibly slow in the limit as the size of the system becomes infinite.

Let us take the exact equation of motion

$$i\hbar \frac{d}{dt} (\mathbf{S}_A - \mathbf{S}_B) = 2J \sum_{\text{pairs}} \left[\mathbf{S}_j \times \mathbf{S}_l\right]$$

The average spins of the two sublattices, $\mathbf{S}_A$ and $\mathbf{S}_B$, are sums of the spins of very many atoms and are therefore rather like classical variables, such as the position and momentum of a macroscopic body; their uncertainties need only be of relative order $N^{-1/2}$. If we suppose that $\mathbf{S}_A$ and $\mathbf{S}_B$ do have macroscopic mean values of some sort, the mean values of the quantities $\mathbf{S}_j$ and $\mathbf{S}_l$, though not their off-diagonal matrix elements—i.e., quantum fluctuations—seem likely by symmetry to be given by

$$\langle \mathbf{S}_j \rangle = \frac{\langle \mathbf{S}_A \rangle}{N} \qquad \langle \mathbf{S}_l \rangle = \frac{\langle \mathbf{S}_B \rangle}{N}$$

Then we may write a c-number equation of motion inserting the *average values* of the variables:

$$i\hbar \frac{d}{dt} \langle \mathbf{S}_A - \mathbf{S}_B \rangle = \frac{2JZ}{N} \langle \mathbf{S}_A \rangle \times \langle \mathbf{S}_B \rangle$$

By setting $\langle \mathbf{S}_A \rangle = -\langle \mathbf{S}_B \rangle$, which is clearly the state of lowest average energy, the right-hand side becomes zero and, at least to order N, the order parameter $\langle \mathbf{S}_A \rangle - \langle \mathbf{S}_B \rangle$ becomes constant. Of course, we have left out the quantum fluctuations on the right-hand side of this equation, but we may suppose that the fluctuations are of order $1/\sqrt{N}$ relative to the mean value of any sum of $\sim N$ elements. This is not at all a rigorous proof but is indicative of the way things go.

On the other hand, because $\mathbf{S}_A - \mathbf{S}_B$ is not a true constant of the motion, its value must fluctuate. In fact, the fluctuation must be of order N. It was proved by Hulthen (103) that the actual exact ground state of any finite system with $\mathcal{H} = J \Sigma_{n.n.} \mathbf{S}_j \cdot \mathbf{S}_l$ (if $J>0$) must have $S_{tot} = S_A + S_B = 0$. This is in fact physically obvious. But in a state with $S_{tot} = 0$, $\langle \mathbf{S}_A \rangle = \langle \mathbf{S}_B \rangle = 0$, because such a state is perfectly isotropic. (This is an exact group-theoretical result.)

This seems to be quite a paradox. Observations and one form of reasoning tell us that $\langle \mathbf{S}_A \rangle = -\langle \mathbf{S}_B \rangle \sim N$ and that this situation does not change macroscopically in time; another reasoning based on symmetry says $\langle \mathbf{S}_A \rangle = \langle \mathbf{S}_B \rangle = 0$. The answer is that the latter is correct in the *exact ground state,* but that this exact state is irrelevant because many other states are infinitesimally close to it ($\sim 1/N$ in energy). Thus if we take as our state one with $\langle \mathbf{S}_A \rangle = \langle -\mathbf{S}_B \rangle \sim N$, this will persist for a very long time but eventually, since it is actually a wave packet composed of many exact states, will break up.

There is a very close and exact analogy with the behavior of a macroscopic solid body in free space. Such a body has a center-of-mass coordinate $\mathbf{Q}$, a corresponding momentum $\mathbf{P}$, and a hamiltonian $P^2/2MN$, N the number of particles. $P=0$ will be the exact ground state, and then $\mathbf{Q}$ will fluctuate wildly: $\mathbf{Q}$ cannot be a constant of the motion. In fact, $\langle Q^2 \rangle \to \infty$ as $P \to 0$. But the levels are enormously close together, so that if we want to localize the whole body within a very small distance, say of the order one lattice constant, $\Delta Q \sim 1$, the corresponding increase in P is ~ 1, in $E \sim 1/N$. Thus the energy levels are so nearly degenerate that we can, for practical purposes, such as for investigating the internal structure of the solid, fix $<Q>$, and then study the internal degrees of freedom before $<Q>$ fluctuates appreciably from its preset value. Perhaps a more direct example would be the orientation θ and angular momentum $I\omega$ of a macroscopic, rigid, rotating body.

But we must realize two things: first, that the existence of the symmetry which makes all values of Q, or of $S_A - S_B$, equivalent, makes the frequency of macroscopic motion of this variable zero. This is exactly the result we had in respect of the order parameter S in ferromagnetism, although in our case the order parameter is not a constant of the motion. The second result is a consequence of the fact that the *fluctuation* of this parameter must be infinite. In the case of short-range forces, the macroscopic energy can only depend on the *gradient* of the order parameter:

$$U \propto |\nabla (S_A - S_B)|^2 \quad \text{or} \quad |\nabla Q|^2$$

But this will mean that

$$<|\nabla (S_A - S_B)|^2> \;=\; k^2 <S_k{}^2> \;\propto\; <V> \;=\; \frac{\hbar\omega_k}{2}$$

so

$$<S_k{}^2> \;=\; \frac{\omega_k}{k^2} \rightarrow \infty \qquad \text{as } k \rightarrow 0$$

We expect, then, $\omega_k \rightarrow 0$ but $\omega_k/k^2 \rightarrow \infty$. Actually, in the best-known cases (antiferromagnetism, phonons in solids and liquids) $\omega_k \propto k$. Surface oscillations of nuclear matter (104) have $\omega_k \propto k^{3/2}$.

The final effect which is always connected with a condensation of this sort, when a low-frequency mode is present, is a long-range force. I will not have time here to be able to go into detail on the inevitable association of long-range forces with zero-mass (i.e., $\omega \rightarrow 0$ as $k \rightarrow 0$) particles, but merely mention the two best-known examples, the Suhl-deGennes long-range nuclear spin-spin interaction (105) via polarization of ferromagnetic or antiferromagnetic spin moments, and of course the well-known long-range character of elastic deformation about a point source which leads to elastic interactions of vacancies, etc., in crystals falling off only as $1/r$.

Now, briefly, I shall give the mathematical treatment of antiferromagnetic spin waves. Incidentally, much of this material on spin waves and magnetism is in my article in Vol. 14 of the Seitzschrift. Antiferromagnetic spin waves are reasonably well covered by Nagamiya, Kubo, and Yosida (101). A very useful trick was suggested by Nagamiya, et al.: namely, to invert the axes for the spins on the B sublattice relative to the A sublattice: for S_l, $y \rightarrow -y$, $z \rightarrow -z$, $x \rightarrow x$, so that

$$S_{lz} \rightarrow -S_{lz}$$

$$S_{lx} \rightarrow S_{lx}$$

$$S_{ly} \rightarrow -S_{ly}$$

therefore,

$$\frac{S_{lx}+iS_{ly}}{\sqrt{2}} = S_l^+ \rightarrow S_l^- \qquad S_l^- \rightarrow S_l^+$$

The reason for switching z and y but not x is to make this a proper rotation, so that the commutation rules are retained for the new axes. In terms of the new axes,

$$\mathcal{H} = -J \sum_{\text{jl neighbors}} (S_{jz}S_{lz} - S_{jx}S_{lx} + S_{jy}S_{ly})$$

Using the approximate expansion

$$S_{jz} = S - \frac{S_j^- S_j^+}{2S}$$

which we already discussed in the ferromagnetic case,

$$\mathcal{H} \simeq -NZJS^2 + \frac{J}{2}\sum_{jl} (S_j^- S_j^+ + S_l^- S_l^+ + S_j^+ S_l^+ + S_j^- S_l^-)$$

When we Fourier transform to spin wave operators (which are now approximately bosons),

$$\beta_\lambda = \frac{1}{\sqrt{NS}} \sum_j e^{i\lambda\cdot j} S_j^{+\cdot}$$

this becomes

$$\mathcal{H} = -NZJS^2 + \frac{JS}{2}\sum_\lambda \left\{ Z(\beta_\lambda^\dagger \beta_\lambda + \beta_{-\lambda}^\dagger \beta_{-\lambda}) + \sum_l e^{i\lambda(l-j)}(\beta_\lambda^\dagger \beta_{-\lambda}^\dagger + \beta_\lambda \beta_{-\lambda}) \right\}$$

This form of hamiltonian is familiar to us from the exciton problem, and if you will look back at the calculation of the "backwards diagrams" in that case, you will discover that the spin-wave frequency is the square root of the *difference* of the squares of the two coefficients:

$$\omega_\lambda = \frac{1}{2}JS \sqrt{Z^2 - \left(\sum_l e^{i\lambda(l-j)}\right)^2}$$

$$\simeq \frac{1}{2}JS \sqrt{Z/3}\ \lambda$$

when the exponential is expanded. Thus we do find verification of our prediction that $\omega_\lambda \sim \lambda$ in this case. We also find that the *zero-point motion* diverges as $\lambda \to 0$; the ordinary amplitudes β_λ may be written in terms of the eigensolutions B_λ as

$$\beta_\lambda = u_\lambda B_\lambda - v_\lambda B_{-\lambda}^{\dagger}, \qquad \text{etc.,}$$

where

$$u_\lambda = \cosh \frac{\theta_\lambda}{2} \qquad v_\lambda = \sinh \frac{\theta_\lambda}{2}$$

and

$$\tanh \theta_\lambda = \frac{1}{Z} \sum_{\substack{l \text{ neighbor} \\ \text{to } j}} e^{i\lambda(l-j)} \to 1 \qquad \text{as } \lambda \to 0$$

so that both u and v diverge. In both two and three dimensions, the divergence in $\Sigma_\lambda \langle \beta_\lambda^2 \rangle$ is integrable, so that one finds merely a certain gain in energy over the nominal energy $-NZJS^2$, and a certain extra zero-point energy of oscillation of the total system about the perfect antiferromagnetic state. We must think of the divergence which occurs in this kind of system as a problem in time scales. The actual $\lambda = 0$ mode does indeed have a divergent amplitude, which is necessary in order that the system be capable of assuming the correctly isotropic ground state; but effectively the frequency of this zero-point motion is negligible as $N \to \infty$, which is to say that the various possible states are "quasi-degenerate," to borrow a phrase of Bogolyubov (105). Thus it is a suitable approximation to form a packet of solutions of the motion of the $\lambda = 0$ mode centered around some pre-fixed orientation, and to study the frequency and the (convergent) zero-point motion of all the other normal modes of spin precession in the time interval before the motion of the packet we have formed breaks up the coherence of the unidirectional state.

In actual fact, of course, anisotropy is always present, leading to a field term which raises the magnitude of the $\beta_\lambda^\dagger \beta_\lambda$ term to (Z + anisotropy) so that the frequency becomes

$$\omega \sim \sqrt{(JZ + H_{an})^2 - (JZ)^2} \sim \sqrt{H_{an} H_{ex}} = H_{ae}$$

the famous "energy gap" expression for the antiferromagnetic resonance frequency. This means actually not that the unidirectional state is really the ground state even in this case, but that the system must tunnel rather than rotate its way around to the time- and spin-reversed state, an even slower process which can certainly be completely neglected. In this system the quasi-degeneracy still exists but only among a finite number of states.

All of the same remarks would apply directly to the system of phonons in a solid. Because the rotational and translational degrees of freedom must have divergent zero-point amplitudes, the frequency goes as $|k|$ and the zero-point amplitude of the lower modes diverges. Similar phenomena are common to He_{II} and to a hypothetically neutral superconductor.

It is important to notice that in general a *continuous* broken symmetry — translation, gauge, rotation — leads to $\omega \to 0$, and usually zero-point amplitude $\to \infty$ as $k \to 0$, where a discrete broken symmetry — such as time reversal in the case of magnets, or interchange of sublattices for order-disorder — need not have any special consequences for the collective modes of the condensed system, unless it happens that there is at least an approximate continuous symmetry, as occurs for the magnetic — and it happens also ferroelectric — cases.

This whole subject of "broken symmetry" is quite new as a formal branch of theoretical physics, although, as I have mentioned, many of the ideas have been understood in particular cases for decades. I do not believe, however, that any general discussion of it from the solid-state point of view exists elsewhere. Still further complications ensue when the interaction of the collective modes of motion with the long-wavelength electromagnetic effects — the Coulomb and long-range dipole forces — is considered. Examples are the magnetostatic effects on spin waves (106) and the plasma effects in superconductivity (107).

BIBLIOGRAPHY

1. C. Kittel, "Introduction to Solid State Physics," 2nd ed., Wiley, New York, 1953.

2. F. Seitz, "Modern Theory of Solids," McGraw-Hill, New York, 1940.

3. J. M. Ziman, "Electrons and Phonons," Clarendon Press, Oxford, 1960.

4. G. H. Wannier, "Elements of Solid State Theory," Cambridge University Press, Cambridge, 1959.

5. R. Peierls, "Quantum Theory of Solids," Clarendon Press, Oxford, 1955.

6. L. D. Landau and E. M. Lifshitz, "Statistical Physics," Pergamon, London, 1958.

6a. M. Lax, "Symmetry Principles in Solid State Physics," to be published (publisher undecided).

7. L. D. Landau, private communication.

8. This discussion is from A. M. Clogston and V. Jaccarino, *Phys. Rev.*, 121, 1357 (1961).

9. F. J. Morin, *Bell System Tech. J.*, 37, 1047 (1958); *Phys. Rev. Letters*, 3, 34 (1959).

10. A good review is: V. B. Compton, T. H. Geballe, and B. T. Matthias, *Rev. Mod. Phys.*, to be published.

11. John R. Reitz, "Methods of the One-Electron Theory of Solids," in Solid State Physics, Advances in Research and Applications, F. Seitz and D. Turnbull (eds.), Academic, New York, 1955, Vol. 1.

12. H. Ehrenreich and M. H. Cohen, *Phys. Rev.*, 115, 786 (1959); for preceding work see references therein.

13. D. J. Thouless and J. G. Valatin, *Phys. Rev. Letters*, 5, 509 (1960).

14. For instance, J. G. Valatin, *Phys. Rev.*, 122, 1012 (1961).

15. G. W. Pratt, Jr., *Phys. Rev.*, 102, 1303 (1956); W. Marshall, *Proc. Phys. Soc. (London)*, 78, 113 (1961).

16. J. C. Slater, *J. Chem. Phys.*, 19, 220 (1951).

17. D. J. Thouless, "The Quantum Mechanics of Many-Body Systems," pp. 24-29, Academic, London, 1961.

18. The first reference I know to this as a general quantum mechanical procedure is F. J. Dyson, *Phys. Rev.*, 90, 994; 91, 1543 (1953). He calls it the "new Tamm-Dancoff" method.

19. F. S. Ham, "The Quantum Defect Method," in Solid State Physics, Advances in Research and Applications, F. Seitz and D. Turnbull (eds.), Academic, New York, 1955, Vol. 1.

20. E. P. Wigner and F. Seitz, "Qualitative Analysis of the Cohesion of Metals," in Solid State Physics, Advances in Research and Applications, F. Seitz and D. Turnbull (eds.), Academic, New York, 1955, Vol. 1.

21. H. Brooks, *Nuovo Cimento Suppl.*, [X] **7**, 165 (1958).

22. H. Jones, "Theory of Brillouin Zones and Electronic States in Crystals," North-Holland, Amsterdam, 1960.

23. The first to consider that nearly free electron theory might be relevant to the diamond lattices was F. Herman, *Phys. Rev.*, 93, 1214 (1954).

24. A. W. Overhauser, *Phys. Rev.*, **128**, 1437 (1962).

25. J. C. Phillips and L. Kleinman, *Phys. Rev.*, **116**, 880 (1959); **118**, 1153 (1960).

26. H. Jones (Ref. 22), Chap. 5, Sec. 44.

27. J. Bardeen, *J. Chem. Phys.*, **6**, 367 (1938).

28. E. P. Wigner and F. Seitz, *Phys. Rev.*, **46**, 509 (1934).

29. E. P. Wigner, *Phys. Rev.*, **46**, 1002; *Trans. Faraday Soc.*, **34**, 678.

30. P. A. M. Dirac, *Proc. Cambridge Phil. Soc.*, **26**, 376 (1930).

31. Probably the best source is D. Pines, in Solid State Physics, Advances in Research and Applications, F. Seitz and D. Turnbull (eds.), Academic, New York, 1955, Vol. 1, p. 367.

32. J. C. Slater, *Phys. Rev.*, **81**, 385 (1951).

33. L. Kleinman and J. C. Phillips, *Phys. Rev.*, **116**, 380 (1959). This is an interesting numerical illustration of many of the ideas of the next chapter. Also note the discussion of exchange in the appendix.

34. C. Herring, *Phys. Rev.*, **57**, 1169 (1940).

35. C. Herring and A. G. Hill, *Phys. Rev.*, **58**, 132 (1940).

36. M. L. Cohen and V. Heine, *Phys. Rev.*, **122**, 1821 (1961).

37. B. J. Austin, V. Heine, and L. J. Sham, *Phys. Rev.*, **127**, 276 (1962).

38. See, for instance, K. A. Brueckner in "The Many-Body Problem," Dunod-Wiley, Paris, New York, 1959.

39. For example: J. M. Ziman, *Phil. Mag.*, **6**, 1013 (1961); C. C. Bradley, T. E. Faber, E. G. Wilson, and J. M. Ziman, *Phil. Mag.*, **7**, 865 (1962).

40. B. T. Matthias, *Phys. Rev.*, **97**, 74 (1955).

41. C. Herring, *J. Appl. Phys.*, *Suppl.*, **31**, 3S (1960).

42. H. A. Bethe and E. Salpeter, "Quantum Mechanics of One- and Two-Electron Systems," Academic, New York, 1957, p. 36.

43. J. C. Slater, *Phys. Rev.*, **36**, 57 (1930).

44. W. Kohn, "Shallow Impurity States in Si and Ge," in Solid State Physics, Advances in Research and Applications, F. Seitz and D. Turnbull (eds.), Academic, New York, 1957, Vol. 5, p. 258.

45. E. I. Blount, "Formalisms of Band Theory," in Solid State Physics, Advances in Research and Applications, F. Seitz and D. Turnbull (eds.), Academic, New York, 1962, Vol. 13, p. 305.

46. Throughout this chapter I shall omit references for the various effects unless they are of particular interest in the exposition, or not well covered in the various texts. Explanations and original references will be found in the texts and reviews referred to. Another good source for free-electron effects is A. B. Pippard, "Experimental Analysis of the Electronic Structure of Metals," *Rept. Progr. Phys.*, **23**, 176 (1960).

47. G. Feher, *Phys. Rev.*, **114**, 1219 (1959). Note Fig. 8; as far as I know, this is the only published version of the theoretical results on the interference effect.

48. G. H. Wannier (Ref. 4), Chap. 6.

49. A. G. Chynoweth, G. H. Wannier, R. A. Logan, and D. E. Thomas, *Phys. Rev. Letters*, **5**, 57 (1960).

50. See especially Pippard, Ref. 46.

51. A good review of the open-orbit situation is given by R. G. Chambers in his article on "Magnetoresistance" in "The Fermi Surface," W. A. Harrison and M. B. Webb (eds.), Wiley, New York, 1960, p. 100. He refers to the important Russian work.

52. J. M. Luttinger, *Phys. Rev.*, **102**, 1030 (1956).

53. R. C. Fletcher, W. A. Yager, and F. R. Merritt, *Phys. Rev.*, **100**, 747 (1955).

54. M. H. Cohen and L. Falicov, *Phys. Rev. Letters*, **7**, 231 (1961).

55. E. I. Blount, *Phys. Rev.*, **126**, 1636 (1962). Another excellent paper giving a more complete theory is A. B. Pippard, *Proc. Roy. Soc. (London)*, **A270**, 1 (1962).

56. References and a more complete discussion will be found in Blount's review (Ref. 45).

57. R. Karplus and J. M. Luttinger, *Phys. Rev.*, **95**, 1154 (1954); J. M. Luttinger, *Phys. Rev.*, **112**, 739 (1958).

58. D. Pines, "The Many-Body Problem," Benjamin, New York, 1961.

59. D. J. Thouless, "Quantum Mechanics of Many-Body Systems," Academic, London, 1961.

60. C. DeWitt (ed.), "The Many-Body Problem," Methuen, London, and Wiley, New York, 1959.

61. V. F. Weisskopf, *Science*, **113**, 101 (1951).

62. R. MacWeeny, *Proc. Roy. Soc. (London)*, **A223**, 63, 306 (1954).

63. L. D. Landau, *J. Phys. U.S.S.R.*, **5**, 71 (1941).

64. Frohlich, Pelzer, and Zienau, *Phil. Mag.*, **41**, 221 (1950).

65. T. D. Lee and D. Pines, *Phys. Rev.*, **88**, 960 (1952).

66. W. Kohn, *Phys. Rev.*, **105**, 509 (1957).

67. J. M. Luttinger, *Phys. Rev.*, **119**, 1153 (1960). See also A. B. Migdal, *J.E.T.P.*, **32**, 399 (1957); *Soviet Phys. (J.E.T.P.)*, **5**, 333 (1957).

68. The development here is close to that of P. Nozières' lecture notes, "Le Problem a N Corps," Dunod, Paris, 1963 (trans.: "Theory of Interacting Fermi Systems," Benjamin, New York, 1963). It is also used in P. Nozières and J. M. Luttinger, *Phys. Rev.*, **127**, 1431 (1962).

69. W. Kohn, *Phys. Rev.*, **110**, 857 (1958); see also V. Ambegaokar, *Phys. Rev.*, **121**, 91 (1961).

70. J. Bardeen and W. Shockley, *Phys. Rev.*, **80**, 72 (1950); for generalization see C. Herring and W. Vogt, *Phys. Rev.*, **101**, 944 (1956); Ziman's book; and Blount's article.

71. This situation was first discussed physically by R. R. Heikes and W. D. Johnston, *J. Chem. Phys.*, **26**, 582 (1957); a more complete theoretical treatment is given by, e.g., T. Holstein, *Ann. Phys.*, **8**, 325, 343 (1959).

72. H. Frohlich, *Phys. Rev.*, **79**, 845 (1950); J. Bardeen, *Phys. Rev.*, **80**, 567 (1950).

73. L. D. Landau, *J.E.T.P.*, **30**, 1058 (1956) [trans.: *Soviet Phys. (J.E.T.P.)*, **3**, 920]. An excellent review of Fermi liquid theory which follows Landau's point of view is A. A. Abrikosov and I. M. Khalatnikov, *Usp. Fiz. Nauk*, **56**, 177 (1958) [trans.: *Soviet Phys. Usp. (London)*, **66**, 68 (1958)].

74. J. Goldstone, *Proc. Roy. Soc. (London)*, **A293**, 267 (1957); W. Kohn and J. M. Luttinger, *Phys. Rev.*, **118**, 41 (1960); J. M. Luttinger and J. C. Ward, *Phys. Rev.*, **118**, 1417 (1960); J. M. Luttinger, *Phys. Rev.*, **119**, 1153 (1960).

75. J. M. Luttinger and P. Nozières, *Phys. Rev.*, **127**, 1425, 1431 (1962).

76. This difficulty is mentioned by Abrikosov and Khalatnikov (Ref. 73), App. 1. Most striking is the failure associated with the B.C.S. theory of superconductivity.

77. J. Frenkel, *Phys. Rev.*, **17**, 17 (1931); *Phys. Z. Sowjetunion*, **8**, 185 (1935).

78. The Wannier exciton seems basically to have been introduced in an excellent paper by J. C. Slater and W. Shockley, *Phys. Rev.*, 50, 705 (1936), although Wannier's paper [G. H. Wannier, *Phys. Rev.*, 52, 191 (1937)] introduces the idea of Wannier functions and the effective hamiltonian as a formal theoretical scheme.

79. J. J. Hopfield, *Phys. Rev.*, 112, 1555 (1958).

80. E. Dresselhaus, *J. Phys. Chem. Solids,* 1, 14 (1956).

81. F. J. Dyson, *Phys. Rev.*, 102, 1217, 1230 (1956).

82. G. Wentzel, *Phys. Rev.*, 108, 1593 (1957).

83. See also T. Matsubara and J. M. Blatt, *Progr. Theoret. Phys.*, 23, 451 (1960), and other papers by the same authors for another approach.

84. M. H. Cohen and F. Keffer, *Phys. Rev.*, 99, 1128 (1955). For the exciton application see W. R. Heller and A. Marcus, *Phys. Rev.*, 84, 809 (1951).

85. R. H. Lyddane, R. G. Sachs, and E. Teller, *Phys. Rev.*, 59, 673 (1941).

86. D. G. Thomas and J. J. Hopfield, *Phys. Rev.*, 124, 657 (1961).

87. N. N. Bogolyubov, *J. Phys. U.S.S.R.*, 9, 23 (1947).

88. For methods similar to these see N. N. Bogolyubov, *J.E.T.P.*, 34, 73 (1958); and Thouless' book.

89. A discussion of this, referring also to Ferrell and others, is given by G. E. Brown and D. J. Thouless, *Physica, Suppl.*, 26, S145 (1960).

90. For these ideas it is better to read one of the books on many-body theory referred to [(58-60); also *Physica, Suppl.*, 26, has a number of good semi-review papers] than to attempt to study the confusing and profuse original literature. The names Brueckner, Watson, and Eden are especially associated with "ladder" summing; Pines, Nozières, Hubbard, and Ferrell with "ring" sums.

91. For a review giving an expanded version of the viewpoint used here, see P. W. Anderson, in Solid State Physics, Advances in Research and Applications, F. Seitz and D. Turnbull (eds.), Academic, New York, 1963, Vol. 14, p. 99; also most of the relevant references may be found there.

92. J. C. Slater, *Phys. Rev.*, 52, 198 (1937).

93. S. Schubin and S. Vonsovsky, *Proc. Roy. Soc. (London),* 145, 159 (1934).

94. J. C. Slater, *Rev. Mod. Phys.*, 25, 199 (1953).

95. C. Herring, *Rev. Mod. Phys.*, 34, 631 (1962).

96. A brief review of spin waves will be found in Ref. 91. More complete is a forthcoming article by F. Keffer in "Magnetism," G. T. Rado and H. Suhl (eds.), Addison-Wesley, Reading, Mass., in press.

97. C. J. Gorter and J. H. Van Vleck, *Phys. Rev.*, **72**, 1128 (1947).

98. Considerations of this sort are discussed by Landau and Lifshitz (Ref. 6), Chap. 13, Sec. 125.

99. M. J. Klein and R. S. Smith, *Phys. Rev.*, 80, 1111 (1951).

100. C. Herring and C. Kittel, *Phys. Rev.*, 81, 869 (1951); L. D. Landau and E. M. Lifshitz, *Phys. Z. Sowjetunion*, 8, 153 (1935).

101. P. W. Anderson, *Phys. Rev.*, 86, 694 (1952); L. Hulthen, *Proc. Roy. Acad. Sci. Amst.*, 39, 190 (1936). A review is T. Nagamiya, R. Kubo, and K. Yosida, *Phil. Mag.*, *Suppl.*, 14, 1 (1955).

102. A name introduced by J. Goldstone, A. Salam, and S. Weinberg, *Phys. Rev.*, 127, 965 (1962).

103. L. Hulthen, *Ark. Mat. Astron. Fys.*, 26A, No. 1 (1938).

104. P. W. Anderson and D. J. Thouless, *Phys. Letters*, 1, 155 (1962).

105. N. N. Bogolyubov, *Physica, Suppl.*, 26, S1 (1960).

106. L. R. Walker, *Phys. Rev.*, 105, 390 (1957).

107. P. W. Anderson, *Phys. Rev.*, 112, 1900 (1958).